A HISTORY OF
THE UNIFORMS OF THE BRITISH ARMY

★ ★

Drummer, "Sergt. XI" Dragoons
1751

Kettle Drummer, 1st Horse
1751

Trumpeter, Life Guards
1751

A HISTORY OF THE UNIFORMS OF THE BRITISH ARMY

BY

CECIL C. P. LAWSON

FROM THE BEGINNINGS TO 1760

VOLUME II

With many drawings by the Author

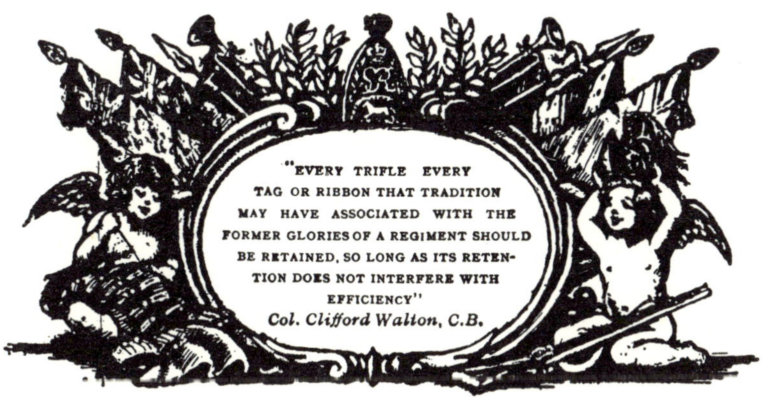

KAYE & WARD · LONDON

in association with Norman Military Publications Limited

First Published by
PETER DAVIES LIMITED
1941

Reprinted by
NORMAN MILITARY PUBLICATIONS LIMITED
1963

Reprinted by
KAYE & WARD LIMITED
194–200 Bishopsgate, London EC2M 4PA
1971

ISBN 0 7182 0201 5

Printed in England by Butler & Tanner Ltd, Frome and London

AUTHOR'S NOTE

In addition to those whose kind help has already been acknowledged in Volume I, I have now to thank the following for their valuable assistance:

The late Mr. A. R. Catley; Major Clarence Goff; Captain Haswell Miller; Major and Mrs. C. M. Vivian Neal; Miss F. S. Philip; Mr. J. O. Rickword; Mr. Kenneth W. Sanderson; Mr. Percy White; Mr. J. B. Wilkie; Miss D. Wilcox.

I am also indebted to the Society of Antiquaries of Scotland for allowing me to reproduce the plate of a 1745 dirk in their collection; and to Messrs. George Waterston & Sons, for permission to copy the dirk and sporans (pp. 67 and 76) from *Ancient Scottish Weapons*, by James Drummond.

Although whenever possible permission has been kindly given by the owners to make drawings from various pictures, in many cases the present ownership and whereabouts of the works are unknown. It is hoped, therefore, that any omission of acknowledgment may be excused.

CECIL C. P. LAWSON

Chelsea
January, 1963

CONTENTS

CONTENTS

LIST OF ILLUSTRATIONS

With Notes on Colours, Sources, etc.

PLATES IN COLOUR

GEORGE I. INFANTRY

AFTER the glorious epoch of the Marlborough campaigns the Government, with the reckless and disastrous policy which has been followed with lamentable regularity to this day, proceeded to disband a large number of the regiments which had done noble service. No consideration was paid to the recognised custom of disbanding the most recently raised regiments first, but old regiments with years of distinguished service were unscrupulously struck off the establishment. To such an alarming degree was the army reduced that it consisted only of between 22—30,000 men, including colonial garrisons, and at the first national danger a certain number of those regiments so rashly disbanded had to be hastily reformed.

On one occasion 25 companies of Chelsea Pensioners were formed for garrison duty to relieve the regular troops. Ten of these companies were regimented under Colonel Fielding as the 41st Foot and still exist as the 1st Battalion of the Welsh Regiment (Fig. 2).

The dress of the Army, with the advent of the Hanoverian dynasty, appears to have been very similar to that worn in the last years of the preceding reign; in fact, as we have mentioned in the previous volume, the dress shown in the mural paintings at Marlborough House was probably that of 1716–20, Laguerre using the soldiers of those days as his models.

There is very little contemporary pictorial representation of the uniform of this time. The wooden figures of Kirk's Grenadiers, till lately at Carlisle, are now in "The Queen's" Museum

FIG. 1. Grenadier, 2nd Foot, 1715.

FIG. 1A.
Cap, loop and details of lace.

(Fig. 1). These are roughly cut out of wooden boards but carefully painted and are over life size, similar to those that were made in France and placed at the barrack gates as a kind of recruiting poster. The paint has blackened very much with age, but since cleaning now shows the lapels a deep blue, three sets of white and yellow loops in pairs, and blue cuffs trimmed broad white lace, with a blue and yellow pattern on either side. The cap is red with a red front embroidered with a coronet above the Prince of Wales's feathers and motto, "Ich dien," either side of this are white scrolls. The little flap is blue, showing a white lamb and round the edge is the motto *Pristinæ virtutis memor* (Fig. 1a). The two Grenadiers appear to be wearing stockings, but this is not clear. The date of these figures is established by the fact that from 1715 to 1726 the regiment was called The Princess of Wales's Own Regiment of Foot, but this was changed in 1727 when the Princess became Queen on the accession of her husband as George II. Two companies of the Regiment were stationed in Carlisle in 1715.

FIG. 2.
The Royal Invalids (41st
Foot), 1751.

Another cut-out wooden figure is that of a grenadier of a Scottish regiment (Fig. 3). The cap is red with a blue front which is embroidered with the St. Andrew's Cross and star with a design of thistles either side, the thistles again appearing on the blue flap of the grenade pouch; this, except in the case of the officers' pouches, seems unlikely unless it is a licence on the part of the painter. The dress is a red coat lined blue and left open at the top with lapels turned back showing the blue lining. The buttonholes have white lace loops and the blue cuffs are ornamented in the same way. The waistcoat is red with loops on one side only; he wears white stockings and not gaiters.

Two little leaden figures (Fig. 4), which formerly stood outside the Chelsea Bun House and now in the National Scottish Naval and Military Museum, Edinburgh, are in a somewhat similar dress except that instead of the St. Andrew's Cross and star and thistle there is one large thistle embroidered on the front of the cap. The little flap is said to have had the white horse of Hanover painted on it, but there was no sign of this when examined in London; otherwise these figures may, in all likelihood, have been of the time of Queen Anne, the white horse having been added on the accession of George I.

A good example of the early Georgian uniform is seen in the Grenadier supporters on the tablet of Sir Charles Wills in Westminster Abbey, which are enamelled in full colours (Fig. 5). They are dressed in plain red coats without lapels but having two rows of lace, looped buttonholes down the chest, the coat is left unbuttoned and displays a red waistcoat. The large buff cuffs are also ornamented with similar loops. The gaiters are white with black buttons.

Fig. 3.
Grenadier of a Scottish Regiment, *c.* 1715.

Fig. 4. Grenadiers, *c.* 1714.

The front of the cap is buff (of a cream colour) and is embroidered with the crest of Sir Charles Wills, colonel of the Buffs, 1716 to 1726, being a black demi-griffin looking to the right, a gilt coronet round the neck, and holding a battle-axe in its claws. The little flap is also buff and has a black grenade with red flames. The cap is very high and more pointed than usually shown.

FIG. 5. Grenadier, 3rd Foot, *c.* 1725

An early Eighteenth Century portrait of an Infantry Officer (Fig. 6) shows him wearing a plain red coat with green slashed cuffs, the cuff being held up by a small chain or gold cord passing over the top button of the slash. The waistcoat is red laced with gold and is of a peculiar cut, the exact details of which are not clear in the painting. A gorget is worn round the neck, the hat is decorated with a fringe of black feathers, and red breeches and white spatterdashes complete the uniform. The arms are the fuzil and sword.

A clothing bill of 1717 for the Coldstream Guards gives the details of the uniforms of their Drummers; scarlet gold-laced coat lined blue, with His Majesty's badges on front and back, cuffs of blue Genoa velvet, blue waistcoat laced with gold, a Mantua silk sash with a gold fringe, red breeches and hat laced gold with black cockade, scarlet cloth surtout, blue facings and lining and cape (collar), with gold buttons and lace.

The Hautboys had the same uniform, but there is no mention of a sash, only the sword and belt. The cost of the latter uniform was £15 13s. 8d., as against £25 10s. for the Drummers, the difference being accounted for by the provision of the gold fringed sash.

FIG. 6. Infantry officer, *c.* 1720.

GEORGE II. INFANTRY

George II on his accession to the throne began to make serious efforts to standardise the dress of the army. It was no doubt with this object that a book was published under official authority in 1742 entitled *A Representation of the Cloathing of His Majesty's Household and all the Forces upon the Establishments of Great Britain and Ireland.* It consists of a series of hand-coloured plates showing a man of all the regiments of Horse, Foot and Dragoons, as well as a Gentleman Pensioner and a Yeoman of the Guard, and was evidently intended to serve as a kind of illustrated Dress Regulations. The work is of extreme rarity, there being only three known complete copies existing in England; a fourth is said to be in America. It is sometimes called the Cumberland Book, as it is suggested that the Duke, who was greatly interested in matters of uniform, was partly responsible for its production.

A Royal Clothing Warrant of 1743 and another of 1749 are frequently quoted in the uniform chapters of regimental histories, etc., but so far the actual Warrants have not come to light. The only evidence of the first is that it appears in a book entitled *The Rudiments of War*, published in 1778. With the exception of one or two items it is practically the same as the Clothing Warrant issued in 1751. It has been suggested that these were drafts for the last-named Warrant. It is possible that its regulations were already being considered or even adopted, as besides these there are endless orders all tending to uniformity throughout the army and curtailing the license hitherto allowed to Colonels in matters appertaining to the clothing of their regiments and especially in the use of their personal crests and coats-of-arms. This is laid down in the words of the Warrant that "No Colonel to put his arms or crest or device or livery on any part of the appointments of his regiment—no part of the clothing or ornament of the regiment to be altered after the following regulations are put into execution but by His Majesty or the Captain-Generals' permission."

Letters now preserved in the Museum of the Worcestershire Regiment give interesting details of the uniform of the 29th Foot during the time of Colonel Desney. They also explain how economies made over one item of dress were used to purchase other articles of equipment.

Gibraltar, 26 Feb. 1727–8: Mention made of patterns of Grenadiers' and Drummers' lace, and "for a grenadiers cap it is not material what you show for a pattern so it be handsome."

Regarding the painting of the drums, it is stated that there will be time before they are made "for Colonel Desney to order his own crest and motto or what other device he likes best."

FIG. 7. Officer, 20th Foot, *c.* 1749.

"Our soldiers coats are faced in the breast and sleeves with a bright yellow cloth, the neck with a yellow cape turned down about two inches; brass buttons and those on the breast smaller than the others; the drums (? frames*) of yellow cloth laced."

Gibraltar, 15 Aug. 1728: "The arms we have are better than any we can get from the Tower with wooden ramrods."

* The spelling or meaning of this word in the original is not clear.

25 Nov. 1728: "An examination of the spare clothing saved here shows 60 good fresh coats and breeches the sleeves of which can easily be altered after the new pattern.*

"The saving of ten suits I would propose as a help towards providing broad-swords for the company of Grenadiers and the additional Grenadiers, for our hangars are most of them broke and the Company when dressed in their new accoutrements look very naked, most of the other regiments have broad-swords.

"It is usual for the Drum Majors coat to be distinguished from the other drums; about which you will ask the Colonel's directions.

"Send also as much lace as will trim the waistcoats of the grenadiers, in place of the coats saved; it need not be so fine as that on the coats, if it be the same pattern."

Gibraltar, 5 Oct. 1729: Black crape for soldiers cravats.

Gibraltar, 9 Feb. 1729–30: "The 360 new shoulder belts and pouches are too good to be used here except on holidays."

Gibraltar 21 Feb. 1729–30: "The Broadswords are come."

It may be as well to state here that, in spite of the fact that as a result of the various orders and warrants the dress of the rank and file was to be brought to a fair state of uniformity, no similar general orders for the dress of officers seem to have been issued. At any rate up to the present none seem to have come to light, no official regulations as to the use of lace to denote rank seem to have existed, and there are portraits of junior officers wearing more or wider lace on their coats than those of senior rank, in fact there still seems to have been a good deal of latitude in the matter.

The portait of WilliamHenry 4th Marquess of Lothian, by David Morier (Fig. 8), shows the full dress of the 1st Foot Guards; a scarlet coat with blue cuffs, lapels and skirt linings,all laced with gold, buff gold-laced waistcoat and buff breeches. The housings, as in his other portrait as Lord Ancrum (Fig. 31), are again buff-laced silver, as would be those of the 11th Dragoons with which Regiment he served at various dates.

By glancing at the accompanying sketches from portraits which have been selected to give some idea of the usual varieties of coat and waistcoat adopted by the commissioned ranks, we will find that there seems to have been a good

* It is possible that this was the cuff and slash as shown on most uniforms of the infantry in the 1742 Clothing Book.

FIG. 8. Officer, 1st Foot Guards, 1741.

deal of latitude still taken by officers in this particular. We find coats laced in various styles, the waistcoats being of red, buff or the facing colour, and either plain or laced on the edges, pocket flaps and buttonholes.

It is true that quite often we find portraits of officers whose uniform is similar to that of the men with the difference of its being laced with gold or silver. In fact one might get the idea that it was largely a matter of regimental or personal taste and that the wearer had more, or less, lace or embroidery put on according to his purse (Figs. 8A, 9, 12 and 16).

Officers are still to be found in coats with red cuffs instead of the facing colour, an example of this being seen in the portrait of Lord Charles Hay of Fontenoy fame, where he is shown wearing a plain crimson red coat worn over a cuirasse. Officers often wore red breeches even though their regiment was dressed in these garments of another colour.

FIG. 8A. Officer, 24th Foot, 1755.

The portrait of an officer of the 24th Foot, painted at Minorca, 1755, is a good example of an officer's uniform. The uniform is scarlet faced with green and laced with silver (Fig. 8a).

A portrait of an officer of the Buffs of the same period shows a uniform coat of the same style and laced in a similar way with silver, except that the pocket flaps of the waistcoat are also laced, the facings and waistcoat, of course, being buff.

The portrait of Captain Baillie in Hove Museum is another example (Fig. 9). Curiously enough, a portrait of Captain Baillie is reproduced in the History of the 13th Regt. of Foot. The two portraits* are identical in every way except that the oil painting has yellow facings and silver lace and the illustration dark buff facings and gold lace. Captain Baillie served in the following regiments: 1744 Ensign 13th Foot; 1747 Lieutenant; 1755 Captain 51st Foot; 1760–1762 Captain 17th Light Dragoons. The portrait with yellow facings and silver lace probably represents him when he was Captain in the 13th Foot. He was, besides a soldier, an artist of considerable talent and gained recognition for his etchings of military subjects.

There is also a portrait of Wolfe as an officer of the 20th Foot. This was painted by Joseph Highmore about 1749–50. In this painting Wolfe wears a red coat with collar and slash of same colour. The lapels and cuffs are yellow laced with silver; the waistcoat appears to be yellow and also laced with silver (Fig. 7).

To return to the 1742 Clothing Book we will find in the series of Infantry regiments that there are still five regiments with single-breasted coats

FIG. 9. Officer, 13th Foot, 1744–47.

* The history of the King's Own Yorkshire Light Infantry, by Col. A. C. Wylly.

without any lapels, and there is also a difference in the shape of the cuffs, some being the plain round shape while others are furnished with a slash and buttons. It seems probable, therefore, that some of the uniforms are of an earlier date and those with the lapels and slashed cuffs are according to the new regulations.

An entry dated November 17th, 1743, states that "all the coats of infantry be lappeled that are not so already, that all be made with the same sort of pocket as the Scotch Fuziliers, viz., in the plaits of the coat according to the Duke's directions signified to me this day. (Signed) Lt.-General Hawley."

An exception must be made to the above as the Highland Regiment and Invalids did not have lapels.

The coat is still large and loose fitting, retaining the same general appearance as before. The cuff, however, as we have already remarked, is now generally furnished with a slash decorated with three or four buttons and stitched or laced buttonholes (Fig. 10). The origin of this was to keep the large cuff from falling down over the hands. In some paintings of officers a small chain or cord is seen attached to the edge of the cuff and looped over the top button of the slash (*vide* Fig. 6).

It will be noticed that most of the regiments now have a lace with coloured stripes or patterns introduced as a means of distinguishing the corps besides its facings.

The Foot Guards, 4th, 8th and 18th Foot, the 7th, 21st and 23rd Fuziliers have blue breeches as being Royal regiments, though strangely enough the 2nd Queens are wearing red like all the remaining regiments. This is probably accounted for by the fact that the men are wearing the red breeches made out of their old coat skirts, about which we will read later on.

FIG. 10. 36th Foot, 1742.

The waistcoat with one or two exceptions is red, which is what one would expect, as the coat after a year's use was cut down so as to serve as the undergarment and having sleeves was used for undress and drill order. The coat skirt linings, or turnbacks, are nearly all of the facing colour. It was now customary to wear the coat skirts looped back when on parade with the result that they hung either side showing the front of the waistcoat and the hindquarters of the wearer; the original idea was no doubt for greater ease in marching, very much in the same way that the French infantry do to this day.

This typical arrangement is so often wrongly shown in modern representations and reconstructions, where the skirts are made to appear to fall close together, as at a much later date, or like the modern tail coat.

A regiment placed forty-third in the Clothing Book and in Millan's list as entitled 61st American Regiment was raised in America in 1739 and consisted of four Battalions, each consisting of a Grenadier and eight Battalion Companies. The first Colonel was Spotswood, who was succeeded by Gooch, both being Governors of Virginia. The regiment served in Jamaica and at Carthegena and was disbanded in 1742. Boyse in his *Historical Review* published in 1748 says "that each man was to have a red camblet coat, brown linen waistcoat and 2 pairs of canvas trousers." The Clothing Book shows a red coat with green lapels and cuffs with white stitched buttonholes. Plain red waistcoat and breeches. The skirts of the coat are lined with a peculiar warm mauve grey, which is probably meant to represent the linen lining which was usual for troops serving in warm climates.

In two of the copies of the Clothing Book appear a series of eight red-coated infantrymen. The first six of these have cuffs, lapels, and skirt linings of a dull almond green. There is no lace on either the cuffs or lapels, the buttonholes being merely stitched with white; plain red waistcoats and red breeches for the whole eight corps (Fig. 75, *see page* 89).

The first and sixth of the series have brown skirt linings in place of green, the sixth having the buttons on the lapels placed in two sets of four. The last two regiments have no lapels, the coats being fastened down the front, the one by a single row of buttons and the other by two rows, set on in pairs with narrow yellow loops, the first having brown linen skirt linings and the latter green.

These plates come at the end of the book after line regiments, and as no number or Colonels' names are shown it is not certain what corps they were supposed to represent, but it now seems highly probable that they are intended for the Independent Companies, which were sent on garrison duty to the Colonies, Companies being stationed in New York, Georgia, Jamaica, Bermuda, etc.

23rd or Royal Welch Fusiliers 1742. 21st or Royal North British Fusiliers 1742.

FIG. 11.

In Millan's List published in 1742–49 these Companies are described as being mostly dressed in red coats faced with green, and it is also mentioned that all regiments serving in the West Indies were to have their coats lined with brown linen. Here we have the explanation of these odd brown skirt turn-backs. It should be noted that West Indies was used in a rather vague and wider sense for overseas stations than at the present time.

The following order gives further details of these companies:

1730, May 30th. "The coats of those Independent Companies such as had wings or lacing on the sleeves were ordered to be taken off and made plain. The facings and lining of the clothing for Garrison or Independent Companies which have usually been blue, we have ordered to be faced and lined with Popinger green instead of blue, after this year with His Majesty's pleasure."

The Pouch was still carried on a shoulder belt and was sometimes buff and at others had the outer flap blackened and decorated with a crown and G.R. The shoulder belt was ordered to be passed under the waist belt in front and over it at the back.

FIG. 12. Officer, Foot Guards, *c.* 1756.

The bayonet and hanger were carried in a wide frog from the waist belt; at the end of the shoulder belt below the buckle can be seen the picker and brush for cleaning out the touch hole of his musket.

The Fusilier Regiments are shown wearing their distinctive head-dress (Fig. 11). This was ordered to be a cap like that of the Grenadiers only lower in height, but in the various paintings there is no appreciable difference, so it seems likely that, as at a later date, it was made to resemble that of the Grenadiers as closely as possible.

The Grenadier Companies are not shown in the 1742 Clothing Book, but the writer has in his collection a coloured 1st edition of the *Grenado Exercise*. These drawings were "drawn from life and etched by Bernard Lens in October 30, 1735," as he states on his title page, for the Duke of Cumberland and consist of 17 plates showing Grenadiers of the 1st Foot Guards in the different positions of the grenade drill. Up to the present this is the only known coloured copy of the 1st edition, the original drawings being in the possession of the Scots Guards. These, on comparison, are identical in colouring with the writer's copy.

The cap is red with a blue front and "little flap" although one would expect this to be red to be heraldically correct, as it bears the White Horse of Hanover. Round the edge of the little flap is embroidered the motto, *Nec aspera Terrent.* On the front of the cap is the Crown, and Star with the Garter, having in its centre the red Cross of St. George on a white field, while issuing from the angles of the cross are narrow dark lines giving the effect of a saltire (Fig. 13).

FIG. 13. Grenadier, 1st Foot Guards, 1735.

There is a curious detail of dress which is not seen on any other figure of this period and that is the cuff, which is decorated with a double zig-zag of lace running round it in addition to the white edging.

In the second edition the first and last figures (which are shown as Corporals in both editions) have the slashed cuff like that shown in the 1742 Clothing Book, the remainder still having the zig-zag pattern.

The second edition was published in 1744. The Corporals in both editions are distinguishable by the white shoulder knot hanging from the right shoulder.

An original water-colour by Paul Sandby at Windsor Castle shows two Grenadiers of the Coldstreams, *c.* 1745, wearing caps like those shown by Lens, and agree in the general style of dress, allowing for the different arrangement of the laces of the two regiments. They appear to be on guard and are wearing white stockings and not spatterdashes (Fig. 14).

Fig. 14. Coldstream Guards, 1745.

It is curious that Lens does not show the match case on the shoulder belt as one would expect.

In Windsor Castle there are two small highly finished paintings of Grenadiers by Lens (Fig. 15). These appear to be about 1735, the same period as his Grenade drill book. The coat skirts are not buttoned back and the edges are bound with lace down the front. They differ in the arrangement of the lace loops on the lapels, these being placed close together and being more in number, and the cuffs are of the slashed variety. These details make one think that these figures represent Grenadiers of the 3rd (or Scots) Foot Guards, as they agree with the drawings of this regiment in the Clothing Book. Their caps show the more pointed shape and the little flap is more rounded in form; otherwise these are decorated, as one might expect, with the same devices, except that the crown is omitted. Here, again, no match case is shown, but that these articles of equipment were worn at this time we know from a reference to the 24th Foot wearing theirs at the back of the belt.

Fig. 15. Grenadier, Foot Guards, c. 1735.

For the Grenadier Companies of the line regiments the 1743 draft says: "The front to be the colour of the facings with King's cypher embroidered with crown over it, the little flap to be red with the white horse and the motto of the regiment over it."

FIG. 16. Officer, 6th Foot, *c.* 1735.

In the 1747 draft it says "the motto is to be *Nec aspera Terrent*, the back part to be red, the turn-up to be the colour of the front, the number of the regiment may be in figures on the middle part behind; the Royal Regiments and 6 old Corps differ from the above"; that is, they had distinguishing devices and therefore not subject to the preceding regulations.

The cypher and crown seems to have been the usual design, although there appears to have been a certain amount of latitude as regards certain details. For example, the little flap, as we have already seen, was certainly blue and not red in several cases, as a letter dated November 17th, 1749, says that "the little flap of Lord Dunmore's (3rd Foot Guards) Grenadiers cap to be red instead of blue according to the new regulation."

It is often questioned if officers of the Grenadier Companies wore the cap. Of this there can be no doubt as we have already seen an instance in a 1st Guards officer in the reign of James II, and there are many caps of this period existing and the fact that they are faced with velvet instead of cloth and the elaborate embroidery being of silver or gilt proves them to be for officers. Again, there are several bills for officers' caps, the cost of which was very heavy, so for this reason it is possible a hat was sometimes worn while on active service.

"The sashes of all officers of Infantry were to be of crimson silk and worn over the right shoulder. Their sword knots to be of crimson and gold in stripes" (Fig. 16).

"Their gorgets either gilt or silver according as the lace of the uniform is either silver or gold. Silver or gold aguillettes were worn as a distinguishing emblem of an officer (Fig. 16). Sergeants are to wear worsted sashes round the waist; they are usually red with a stripe of the colour of their facings" (Fig. 30).

Corporals wore a white worsted shoulder knot (Figs. 32 and 46).

The men were furnished with brown marching gaiters either canvas or cloth with black buttons, besides the white ones for full dress. According

FIG. 17. Pioneer, Foot Guards 1745.

to the orders of the Foot Guards these "brown cloth gaiters were only to be worn on detachments and out parties and when the men wore them the officers were to wear boots."

In Hogarth's picture of the March of the Guards to Finchley we find many details of uniform not shown elsewhere. One figure represents a Pioneer dressed like the rest of the Battalion but wearing a cap with a loose hanging red bag terminated with a white tassel and turned up all round with a deep band of blue on the front of which, embroidered in white, are a mattock and saw (Fig. 17). The back has the usual scroll embroidery also in white. He wears the Infantry equipment of waist and shoulder belts and pouch, but as his left side is hidden one cannot tell if he wears any side arms. No leather apron is shown, but it is probable that he would have had one seeing that they were already used by Farriers and in 1755 were issued to the 15th Foot.

In the same painting a Sergeant is shown, unfortunately only his head and shoulders are to be seen, the rest being hidden by other figures, but even so we can see that the blue lapels of his coat are edged and looped with gold lace and that he carried a halberd.

Until 1725 Corporals of the Foot Guards had been armed with firelocks, but in January of this year they were ordered to carry halberds in their place. The officers' fire-arm would be the fuzil and Company officers would carry an espontoon besides their swords (Fig. 37).

In 1737 the officers of the Coldstream Guards had blue frocks. So far as is known at present there is no picture which shows this coat, unless the portrait of the Duke of Cumberland represents him wearing one of these garments. If so, it was a plain blue coat with lapels and cuffs of the same colour edged with gold lace, the buttonholes being quite plain. The Duke's waistcoat is red edged with gold lace. It seems that it was by no means unusual for officers to wear blue undress frocks, as in 1756–57 we hear that deserters from the 40th and 62nd (Royal Americans) "took away with them blue surtout coats," and again we read that officers of Militia had blue frocks faced with red. Besides these there are numerous other mentions of these blue undress coats.

The writer was very kindly allowed to see the Order Books of the Grenadier Guards, which contain much valuable information concerning dress, and from which the following extracts are quoted. On the death of Queen Caroline in 1737 the following orders for mourning were issued to the Foot Guards: "Every officer to have a scarlet coat buttoned to the waist with a mourning button and faced with black cloth, no buttons on the sleeves or pockets, black cloth breeches and waistcoats. Plain hats with crêpe hat bands, an end appearing at each corner of the buttoned side of the hat, mourning swords and buckles, and to get crêpe for their sashes."

"1748, 25th April, officers of Grenadiers to wear their sashes under their coats and round the waist and sword belts under their waistcoats."

"1749, 20th March, officers to make up buff coloured cloth waistcoats, laced with the same gold binding as that the regiment's frocks are bound with" (Fig. 12).

The officers of the Coldstream Guards the same year, when on duty, were to wear buff breeches as well as waistcoats of this colour.

The Company officers of the 1st Guards (May 31st, 1749) are "to see that the men's new clothing fits well, particularly the coat is not too long. The seam at the bottom of the waistcoat to be well strengthened with a loop to prevent ripping."

To fit a man in his coat the following rule to be carefully observed throughout the regiment—"that when the man is kneeling the bottom of the skirt of his coat to be two inches from touching the ground and hang an equable length all round. The shoulder strap to be placed high on the shoulder and the end well fastened down and turned under."

"The lace of the waistcoat to be sewed down in the middle as that on the coat and care to be taken that the loops on each side of the waistcoat answer exactly to one another and to be of equal length. Every man to have an oyle picker and brush and match cases to be fitted to the belts of the Grenadiers.

"No soldier in new or old regimentals with lace on to carry any burthen, bundles, wear an apron, drive carts, or do any kind of work, never to appear in the street without his sword on, hat well cocked, and hair tucked under."

The punishment for failing to obey the above orders was "to be sent to the black hole with bread and water for 40 hours."

We have already seen that the old coat was cut down to serve as a waistcoat, but besides this the Orders of the 1st and 2nd Foot Guards state that the cloth from the skirts were to be made into a pair of breeches. These were to be for undress, the blue being kept for full dress. One Order of the Coldstreamers says that the men were "to have good blue breeches on because their clothes are to be looped up."

Officers at first had perukes somewhat similar to those worn in Queen Anne's time, but smaller and powdered. According to the Orders of the Coldstream Guards for a review in 1735 the officers of the regiment "are to have on twisted Ramilyed wigs." In 1747 they "are to wear queued wigs or their own hair done in the same manner." The officer of the Guards in Hogarth's picture has a long thin pigtail. This seems occasionally to have been worn by officers but not by the men.

From the appearance of the Privates in the 1742 Clothing Book, it is often thought that their hair was worn cut short, but from many references of the time it is clear it was worn tightly brushed up and tucked under the hat.

The Guardsman (Fig. 18) has his hair neatly tied in a very short queue. This print is one of the very few which show how the soldiers' hair was worn at this time and is definite proof that it was not cut short.

FIG. 18. 1st Foot Guards, 1745.

Wigs were not allowed to be worn "except in the case of those men who cannot wear their own hair through age or infirmity, and then they are to be so like a head of hair as not to be perceived and to turn them up like their own hair which they are to wear on Mounting days."

Morier's paintings show the Grenadiers with their hair plaited and turned up at the back and tucked under the cap. The Grenadier of the 36th Regiment, however, is shown with long hair hanging down as in the last days of the seventeenth century (Fig. 38). The hair is worn in a sort of curl on each side of the head above the ears.

In the Royal United Service Institution is an early Grenadier cap of the 67th Regiment. This appears to have had the front stitched back to the upright cloth crown, giving it a conical shape mentioned before, but this may have been done at a later date as a means of preservation. The curious point about this cap is it has a long red horse-hair fringe knotted along the base at the back; under this again is a pointed strip of horse or some other animal's skin about two or three inches long with the hair still on. This is stitched just inside the base of the cap. The whole arrangement appears to have been some kind of false peruke like those mentioned in the foregoing Orders, though why it should be the same red coloured horse hair as the tuft or tassel of the cap is a mystery. It does not appear to have been caused by age or by being dyed at a later date. The cap itself, which is 10¾ inches high, is red with the front and turn-up at the back of the yellow facing colour; the little flap is red (Fig. 18a).

The following references to dress of the Foot Guards may also be added:

"Volunteers from the Foot Guards for Flanders distinguished by black and green cockade." (The *Ipswich Journal*, 9th May, 1747.)

1st Regiment of Foot Guards reviewed by the Duke of Cumberland: "The officers of the Grenadiers had caps, belts and cartouches with fuzees, agreable to the custom of these corps abroad." (The *Gentleman's Magazine*, May 1749.)

"We hear that an alteration will be made next summer in the full regimentals worn by the officers in the Foot Guards by adding embroidery to their coats." (The *Ipswich Journal*, 3rd March, 1753.)

FIG. 18A. Cap with wig attached, 67th Foot.

5th Regt. 6th Regt.

FIG. 19. Marines, 1742, Colours, 44th Regt., *c.* 1750.

MARINE REGIMENTS

1739–1748

IN 1739 six Marine Regiments were ordered to be raised, the 1st or 44th of the Line; 2nd or 45th; 3rd or 46th; 4th or 47th; 5th or 48th; 6th or 49th. N.C.O.'s were drawn from the Foot Guards and a bounty of 30s. was offered to the first 1,800 men who volunteered from the line regiments.

MARINE REGIMENTS

Four more regiments were raised in 1740.

The uniform was to be "a good full-bodied coat, with red waistcoat made from last year's old coat, red Kersey breeches, long brown canvas gaiters." All these regiments are shown in the 1742 Clothing Book, and it will be noticed that there the 6th and 9th have gaiters of blue and white striped ticken, all have grenadier caps, the fronts, however, being of a different shape to those of other regiments. The front and little flap are both of the facing colour and are all edged with red lace. The device is the same for all regiments. viz., a crown over G.R. embroidered in red except the 4th, 7th and 9th, where it is yellow. The little flap has the garter star (Fig. 19).

1st Regiment yellow cuffs and lapels.

2nd Regiment green cuffs and lapels.

3rd Regiment deep yellow cuffs and lapels.

4th Regiment white cuffs and lapels and stitched buttonholes.

The lace for this regiment was a red and yellow stripe with a white and blue edge.

5th Regiment pale yellow cuffs and lapels and pockets and lapels bound with white lace (Fig. 19).

7th Regiment white cuffs and lapels, which appear to be much shorter and have the buttons and stitched buttonholes in two sets of threes.

8th Regiment bright yellow facings.

9th Regiment dark buff, almost brownish, facings; this tone may be caused by the oxidizing of the colours.

10th Regiment deep buff facings.

The 6th has no lapels, but three double sets of buttons and buttonholes and a green turn-down collar (Fig. 19).

The buttons and lace for every regiment are shown white except that of the 4th Regiment.

There has been some question as to what the back of the cap was like, but there appears to be no reason to think that it was different to the usual grenadier cap except possibly the crown was not stiffened but hung down at the back like those of drummers of the same period.

Their equipment consists of a dark buff waist belt and two black pouches; no shoulder belt is shown as for line regiments.

In 1747, May 5th–9th, the clothing of the ten regiments was ordered to be as follows:

For a Private "red coat to be breasted in the lapel fashion, face turned up with bright yellow, white metal buttons, good Kersey breeches, stockings, shoes, etc. A cap with an anchor and cable embroidered on the front. Corporals the same with in addition a white shoulder knot. Sergeants the same as the privates, except the breeches to be of cloth instead of Kersey and the whole a better kind. Drummers the same as the private except the coat is to be of yellow cloth with a proper lace thereon."

The ten regiments were disbanded in 1748.

The position of the Marines is somewhat puzzling, as they seem to have been partly under the Admiralty and yet appear in the Army Lists numbered according to their sequence in the line. Their actual position was also very ambiguous, especially when the time came to pay the regiments, when both the Admiralty and the War Office declined all responsibility in the matter. Finally they were definitely placed under the Admiralty in 1747.

FIG. 20.
Wolfe as volunteer in the Marines.

The colours of the Marines are shown elsewhere, only those of the 1st Regiment (or 44th Foot) are depicted in the Windsor Colour Book, but it was ordered that all regiments of Marines were to have in the centre of their colours a ship with sails furled and the number of the regiment underneath; otherwise they would no doubt follow the same rules as the line regiments. Their caps, bells of arms and drums were to follow the general regulations.

A portrait of Wolfe as a gentleman volunteer in the Marines shows him in a single-breasted coat with laced button-holes arranged in an oval loop design (Fig. 20).

Another portrait of Wolfe in the National Portrait Gallery shows him at a later date also in the uniform of the Marines. The coat is furnished with yellow lapels, cuffs and waistcoat edged with white (or silver) lace. It has been suggested that this is the portrait of his brother at that time, Captain in the 1st Regiment of Marines (Fig. 21).

all edged with white. The front is embroidered with a peacock in yellow (or gold) and blue standing on a coronet with branches of laurels either side, the whole surmounted by a crown, the little flap has a white edge piped with red and bearing the motto *Pour Y Parvenir* in black letters. The flap itself has the same design of trophies of flags as the Harcourt cap. The blue turn-up is edged with white and embroidered with laurel leaves. The cap is piped with white and has a red and white tuft. This regiment was raised by the Duke of Rutland, October 4th, 1745, the Colonel being the Marquis of Granby, who was 24 years old at the time. The regiment had a strength of 780 men in ten companies, its Lt.-Colonel being Stanwix, who later commanded the 8th Foot. The regiment took part in the pursuit of the Jacobite forces as far as Newcastle. The following arms were issued to the regiment: 300 carbines new pattern, 402 pairs of pistols with carbine bores, 202 broad swords, 12 drums, cartouche boxes and belts. The number of pistols issued to an infantry regiment is curious. The arms and tents were issued by the Ordnance to all these regiments.

A portrait of Granby by Sir Joshua Reynolds painted in 1747 is said to be the earliest portrait of Granby and to show him at the age of 26, wearing the uniform of this regiment. If this be correct the uniform was blue with red lapels and cuffs and red waistcoat. All are edged with gold lace and have gold-laced buttonholes, being, in fact, very like the uniform of the Horse Guards which he joined later. It certainly seems unusual for a regiment of foot to be wearing a blue coat at this period unless it is the undress blue surtout. The grenadier cap is certainly red and blue, but one would have expected the coat to be red faced blue and not blue faced red. This portrait originally belonged to his brother Lord George Manners-Sutton and later to Herbert A. Sutton, Esq., of Kelham Hall, Newark, and was at one time in the Sedelmeyer Collection which was sold in Paris in 1907 (Fig. 23).

FIG. 23. Officer, 1745.

1750

We have already noted the desire of George II to regularise the dress of the army and in this he was largely assisted by the Duke of Cumberland. With this aim the various drafts and dress orders were now embodied in the Clothing Warrant of 1751 (which is printed in full in the Appendices). In it are laid down the regulations for the dress of officers, N.C.O.'s, men, drummers, etc. Besides the Warrant there are the paintings by Morier in Windsor Castle, which were made about the same time. David Morier was of Swiss origin, and was taken under the patronage of the Duke of Cumberland, from whom he received an annuity and who commissioned him to make this series of oil paintings representing a soldier of every regiment of the regular army (Fig. 24).

It has been pointed out by several experts of uniform that Morier could not have had a soldier from every regiment for his model owing to the unit being abroad or on a distant station. It seems most probable, however, that the Duke intended these paintings to serve as a guide or pattern as to how each regiment ought to have been dressed according to the regulations contained in the Warrant, and as such they form the principal and most priceless source of information for this period.

Fig. 24. Grenadier, 8th Foot, The King's Regt., 1751.

It is possible that Morier had a man dressed in the sealed pattern coat of each regiment; this is rather supported by the fact that in most cases there is a strong resemblance in the faces of his figures.

The Infantry is represented by a grenadier of each regiment (Fig. 33) and we shall see by comparison with specimens of cloth caps still existing and regimental orders and inspection returns that in many cases they agree to a very large extent. The backgrounds of these paintings show camp scenes and many minor details of military interest not to be found elsewhere.

About this date Generals made periodical inspections of regiments and their reports or inspection returns give much information of undeniable authenticity.

With the three foregoing sources we are at last on firmer ground and have a clearer picture of the dress of the army and of the regimental details which distinguished one corps from another. From Morier's paintings we see that there was a similarity in the cut of the coat and general appearance of the dress. The cap is red with a front of the facing colour, lined at the back with red. The little flap is red, having a border of the same colour, but edged with white and having the motto *Nec Aspera Terrent* in white letters. The turn-up at the back of the cap was the colour of the facings, and having a grenade embroidered in the centre and the number of the regiment with ornamental scrolls, or foliage, on either side. The design on the front of the cap is the Royal cypher G R surmounted by the crown and having a design of scrolls, or leaves, on either side (Fig. 25).

The Foot Guards and what are called in the Warrant the Six Old Corps must be excepted as these had special badges which were embroidered in place of the Royal cypher.

The caps of the three regiments of Foot Guards had the blue fronts embroidered with star, etc., exactly like the grenadier cap of 1735, except

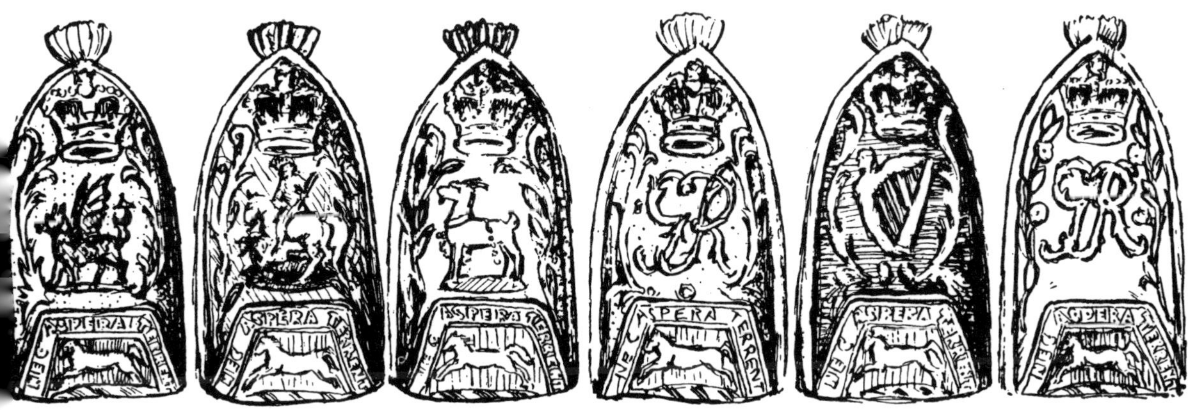

3rd 5th 6th 9th 18th 32nd

FIG. 25. Types of Grenadier Caps, 1751.

that the star has more and smaller rays (*vide* Fig. 32). The little flap, however, had now been changed to red. The majority of the caps in Morier's paintings (excepting those with Royal or special badges) had the G R and scroll design in white. This seems to have been the general design, a crown proper always being placed above. The following regiments, however, had the G R embroidered in coloured worsted, the scrolls remaining white: the 16th in red, 28th, 29th, 34th and 43rd in dark blue; 30th, 37th, 38th and 48th in green; 40th, 44th and 47th in black.

The following regiments, however, had the G R and leaves in green: 24th white leaves; 32nd G R in red and leaves in green.

A. The 3rd Foot. The Buffs. B. The 65th Foot.

FIG. 26. Officers' Grenadier Caps.

Tufts of Grenadier caps, from Morier's Paintings

White with red centre—1st, 2nd and 3rd Foot Guards.

White—1st, 5th, 15th, 20th, 23rd, 25th, 26th, 27th, 31st, 34th and 48th Foot.

Red, white and green—2nd Foot.

Red, white and buff—3rd, 14th Foot.

Blue and white—4th, 7th, 8th, 18th, 21st and 43rd Foot.

Yellow and white—6th, 9th, 10th, 12th, 13th, 16th, 28th, 29th, 30th and 44th Foot.

FIG. 29. Grenadiers in Marching Order, 1751.

46th 27th 49th and 48th Foot.

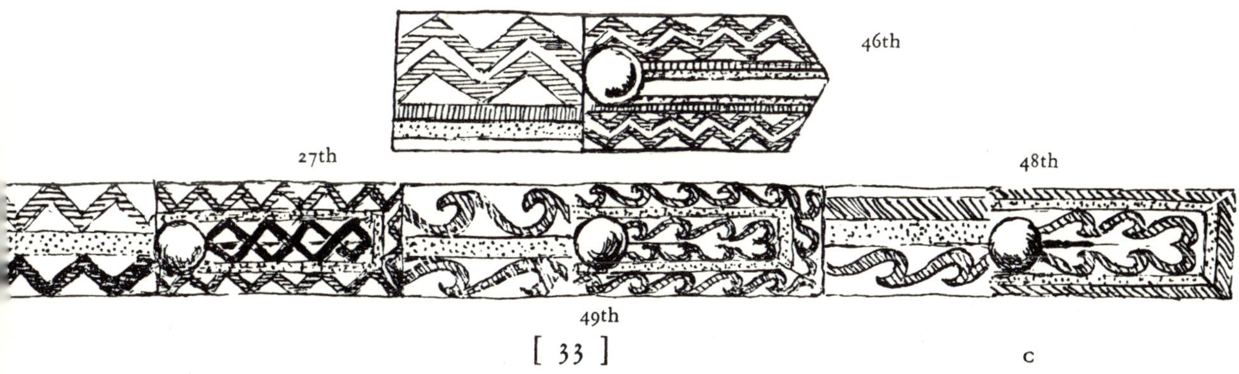

It may be remarked in the foregoing examples that whereas the officers' caps with their elaborate decoration differed somewhat from the Morier paintings, those of the rank and file agree with them quite closely in their design. While on the subject of caps, we find that in 1755 the grenadiers of the 30th Regiment wore bearskin caps, their drummers having them of white fur. From this it appears that although the fur cap was not introduced officially until the 1768 Warrant, regiments had already begun to adopt them for their grenadiers before this date. As far as one knows this is the first instance of a grenadier rur cap for this period, always excepting the Highland regiments for whom it was the regulation head-dress.

The cuffs, lapels and skirt linings were nearly always of the facing colour, the 24th Foot having green facings and white linings.

On some of Morier's figures the coat is seen buttoned across the chest to the waist and showing the lapels turned back at the top, the waist-belt being worn outside the coat; this was for marching order. On some of the other figures the coat is shown left open in front with the lapels buttoned back to their full length, the waistcoat being thus fully exposed, the waist-belt in this case being worn under the coat. The neck of the coat has an edging of lace which, at times, forms a sort of low collar. The cuff slash is invariably red with a lace edging and buttonholes. The skirts are ornamented with various styles of lacing and pocket flaps. Every coat has a plain red shoulder strap on the left shoulder except one, and some regiments wear wings ornamented with lace. The waistcoat is red and furnished with two pockets; it is generally bound down the nexk, front and skirts with lace and having the buttonholes sometimes laced or stitched with white.

The grenade pouch is still carried suspended to a wide shoulder belt to which the match case is attached. It had now become customary to have the pouch flap blacked, the Guards still having the G R and crown in brass on theirs.

The sword and bayonet were carried as before on the waist belt, over which was buckled a strap carrying the cartridge box.

The sword hilts vary considerably in pattern, but generally are of the basket hilted type with a guard of pierced iron. In the Inspection Returns we read of swords of different patterns for the Grenadier and Battalion companies, the first apparently having a hanger with a curved blade and the latter a straight one. In other cases we find that a regiment is without swords, except the Grenadier company.

FIG. 30. 4th or King's Own Regiment, 1743.
Officer, Sergeant and Privates of Grenadiers and Officer "Bn" Company.

In marching order the men are shown with a cowhide knapsack and water-bottle or flask of white metal on the left side, the haversack of greyish-coloured canvas on the right. One Grenadier is painted wearing his waist belt over the right shoulder (Fig. 29).

Captain John Knox of the 43rd Foot mentions an incident of an English ship being mistaken for a French transport bound for Louisbourg because "our soldiers had white linings to our uniform and the cloathes being turned outside in, for cleanliness *according to the custom of troops at sea*."

The men of the 43rd Regiment thus appeared to be dressed in white coats, similar at a distance to those worn by most of the French Infantry of that time.

This practical custom of wearing the uniform coats turned inside out must have produced a curious effect in the case of a regiment having yellow, bright green, orange linings, or sky-blue linings.

FIG. 31. Officer, 24th Foot, 1748–52.

Morier's painting of the Battle of Culloden shows quite clearly the various types in one regiment—the 4th or King's Own. An officer is in a red coat with blue lapels edged with silver lace, the buttonholes, however, being plain, round blue cuffs with two bands of lace, silver aiguillette and hat lace, a crimson sash worn over the right shoulder; his breeches are of red cloth, while those of the regiment are blue. By his position at the head of the Grenadiers and his handling a fuzil he is evidently intended to be an officer of that company.

The Grenadier Sergeant has his coat laced with narrow silver loops in pairs in a different style to that of the men, his coat having plain small blue

FIG. 32. Foot Guards, 1751, Grenadier of 2nd, Grenadier Corporal of 3rd and Bn. Corporal of 1st Regt.

cuffs and no wings. He wears a sash of white, blue and crimson round his waist and is armed with a halbert. Some of the Grenadiers have an unshaven appearance, one even to the extent of having a stubbly beard and moustache. Behind the Grenadiers comes a battalion Company Officer dressed as the first, but carrying a spontoon. Both the officers wear their hair in pigtails, the grenadiers having theirs plaited and turned up under their caps (Fig. 30).

On February 4th, 1752, the Duke of Cumberland ordered that "the Majors and Adjutants of infantry (whenever they appear on horseback) are to wear an uniform horse furniture, which is to be of cloth the colour of the facings of their respective regiments and laced with gold or silver, as the officers cloathes are laced. The Majors to have two laces and a tassel on their housings and (holster) caps, the Adjutants to have no tassel." These tassels, which hung from knotted loops from the centre of the holster and the corners of the housings, are often seen in paintings and on some specimens of horse furniture still in existence.

A portrait of William Earl of Ancram shows him as Colonel of the 24th Foot, which rank he held from 1748 to 1752. Unfortunately his yellowish buff and silver-laced horse furniture does not correspond with the green facings of his uniform; only the holster caps are visible and these are ornamented with a device in silver, the housings being covered by the skirts of the coat. Although the uniform is laced with silver the sash is of crimson silk and gold (Fig. 31).

A further reference to the dress Orders of the 1st Foot Guards will give us not only more details of their own uniform but also the regulation which, no doubt, their Colonel, the Duke of Cumberland, wished to see carried out by the army as a whole.

In 1750 the rules concerning the fitting of the coat were as before except that "the six points of the lapells to be tacked down in such a manner as to be easily unsewed again when required, without damaging the cloth, the bottom of the lapell where it joins the coat on the belly to be well stitched to prevent ripping. The back pleat to be sewed down as low as the lower loop."

In 1751 the coat is made two inches shorter than before. "Regard to be had to the size of the men as to the number of loops to be on each (coat) but the length of the loop and the distance between them to be alike viz: the length 2 inches ⅝ and the distance between each loop ¼ inch. The following

FIG. 33.

17th 16th Sutleress 21st or Royal North British Fuzeliers.

16th. 17th. 21st.

rule to be observed as to the number. The waistcoats for men under 5 foot 8 inches high to have 9 loops only on each side, those of men 5 ft. 9 and under to have 10 and those above 5 ft. 9 to have eleven, the Grenadiers of 5 ft. 11 and upwards to 12 loops."

"The waistcoats to be cut square at the bottom and open back from the lower button hole to the point; the back seam to be sewed down as low as the bottom of the waistband of the breeches, the red cloth of each company to be made up into breeches and delivered out to the men who are constantly to wear them except on such days as they are under arms."

1750—Officers to have gold hat bands: 1753, to have narrow gold binding to their regimental frocks, the lapells to be cut square and the waistcoat to be laced agreeable to the coat, without any lace under the pocket.

1751—The Sergeants' waistcoats are all to be laced with the same as that on the coat.

Officers to wear frocks and light boots when the men are in brown gaiters.

1750—Cap for grenadiers red turned up with blue, a blue front and blue tassels.

Garters (for the gaiters) are to be cut off so as to exceed the buckle by 1 inch and no more.

In 1756 the brown marching gaiters were in some regiments replaced by black ones, the Coldstream Guards ordering that theirs should be blackened and tops put on them. This sounds as if it were a stiff leather top, which we will come across further on.

Fig. 34. Sutleress.

In one group of the Morier paintings a sutleress is shown wearing a bright green jacket, red waistcoat, with white lace loops, brown skirt, light blue apron, blue bow and stockings and low shoes with large red rosettes. The hat is of a surprising bright green colour with yellow lace edging and band (Fig. 34). The semi-military effect of this costume was due to the use of portions of discarded soldiers' clothing. This is the earliest representation of a woman in soldier's garments although later there are numerous examples of it. The practice of the women who followed the army making good the deficiencies of clothing during a campaign by using soldiers' old coats, etc., was no doubt the origin of the fanciful military uniform adopted by the cantinières of the French armies which reached its greatest *éclat* in the Second Empire.

Sutlers played an important part in the old armies as they were dependent

FIG. 35. Sutler's Tent, 1730–40.

on these private speculators for the supply of extras in the way of provisions and liquor. Fortescue says that the only ration supplied by the Commissariat was bread. Sutlers were officially recognized, one being allowed per Company of Foot or Troop of Cavalry, receiving an allowance of forage from the Commissariat for their horses. The exploits of Mrs. Christian Davies (Fig. 34), or Mother Ross as she was later known in the army, explain that a little careful plundering and marauding were useful in supplementing the stock in trade. From the results of some of these activities she was tactful in providing good meals, when food was scarce, to a hungry General Staff, who were equally politic in not inquiring as to the source of supplies. No compensation could be claimed by a sutler for loss of his goods due to the exigencies of a campaign.

On the death of Marlborough his careful supervision ceased and the control of the Board of General Officers seems to have been relaxed. Contracts were no longer examined, the certificates were no longer lodged with the Judge Advocate before regiments were fitted out as had always been the rule and, in fact, the Colonel once more had a free hand without any check on the prices, quantities or quality. More attention was now paid to the soldiers' uniforms being made according to dress regulations than to its serviceable qualities. The cloth was supposed to be dipped in water and shrunk before being made up. This some Colonels refused to do on the pretext that it spoilt the look of the uniform, with the consequence that when the men had been soaked by rain their uniforms became so tight as to hamper their movements.

FIG. 36.
Corporal, 13th Foot, 1747.

Steel ramrods, invented, it is said, by a German officer who had served in Marlborough's army, began to be issued in 1725 in place of the wooden ones hitherto in use, although it was some time before they were universally adopted.

To save the flints from being worn out in musket drill wooden or bone snappers were fitted in their place.

Canvas covers were provided to protect the musket locks and priming, and seem to have been chiefly used by the Cavalry. The tents were the ridge pole type, as seen in Morier's paintings. These held six men, and a camp kettle with three feet was given to each tent. Bell tents, or Bells of arms, which are mentioned in the 1751 Warrant, were used for storing the arms when in the field.

The Duke of Cumberland was a strong, able administrator and a strict disciplinarian. In spite of Walpole's jibe "that he was as intent on establishing the form of spatterdashes and cockades as on taking a town," it was due to his work and energy that the army was retrieved from the chaotic state of disorder and indiscipline to which it had been reduced by carping and factious politicians after the guiding hand of Marlborough had been removed.

If he was pedantic over details of military clothing, it seems to have been with a view to getting regiments uniformly dressed, and his order that officers under the rank of Brigadier should always appear in their regimental coat, even if an old one, was more with a disciplinary than a sartorial end, as was the order that wheeled vehicles should be limited to two per regiment, one for the Colonel and one for the Sutler.

Fig. 37.
Officer's Espontoon.

It has frequently been asserted that by his influence as well as the King's the men's uniforms were made tight fitting in imitation of the Prussians. This statement has so often been repeated that it is now accepted without question. It would be as well, however, to examine the facts. The uniforms of the Prussian infantry during the first part of Frederick II's reign resembled those worn by other European armies, being just as easy fitting, if perhaps a trifle shorter. The source of this legendary Prussian tight uniform seems to have originated with Adolph Menzel's drawings of the soldiers of Frederick's army. These, however, although based on actual uniforms, represent the shorter and close fitting uniforms adopted in 1786 at the end of Frederick's reign and consequently differ very considerably from those worn during his campaigns.

A glance at the 1742 Clothing Book and the pictures of Morier and other artists of the period, or even those of a later date, shows no evidence of the so-called tight Prussian style. In fact it seems to be part of the legend which attributes all military innovations to the Prussian King. His success at Rosbach has been largely attributed to his infantry having steel ramrods and the French wooden ones, whereas the latter were equipped with them as well.

It has already been shown that it was not until 1742–3 that Frederick started training his cavalry in shock tactics, which had been in use since the days of Cromwell and Rupert and improved by Marlborough. C. T. Atkinson in his *Marlborough and the Rise of the British Army* points out that fire discipline and control had already been the distinctive feature of Marlborough's infantry; "and that except for the more highly developed and flexible drill and consequent increase of manœuvring power there is little in Frederick's tactics that is not found in Marlborough's, and that Ramilies was an anticipation in principle, if not in detail, of the oblique order of Frederick."

Fortescue says that he can find no evidence for the statement that the excellence of the British fire at Fontenoy and other battles was due to the adoption of the Prussian drill, which indeed found its way to the 1st Guards in 1756—years after Fontenoy—but had by no means been universally adopted.

FIG. 38. Grenadiers of 36th, 34th, and 23rd Foot, Royal Welch Fuziliers, 1751.

That the Prussian drill and precision of march, learned with pendulum and pace stick, was in advance of other nations and was to have a lasting influence is not disputed.

Both the above authorities trace the origin of the practical improvement in tactics to those British officers who served and learnt under such masters as Maurice of Nassau, Gustavus Adolphus, Turenne and others.

It had now become the practice to unite the Grenadier Companies of the various regiments in the field into one body, an example being the Louisbourg Grenadiers, so called from being composed of the Companies drawn from the 22nd, 40th and 45th Foot stationed at Louisbourg. This practice was deplored by some officers, on the grounds that the Grenadiers being selected from the tallest and best men of the battalions it was supposed to reduce the value of the remaining Companies. Maréchal de Saxe in *Les Reveries* inveighs against the perpetual use of Grenadiers which, in a serious war, so reduced their numbers that they are no longer able to supply the N.C.O.'s which are the backbone of the infantry. He continues: "The custom of detaching grenadiers on every service even the most trifling has been the cause of the lives of brave men being too often sacrificed uselessly."

Europe, however, was not the only battleground of the army, as America was now to become an important field of operations which were to have their influence on dress and tactics.

The unfortunate General Braddock, notwithstanding what has been said, seems to have made a genuine attempt to adapt the equipment of his men to the novel conditions of forest fighting as his following orders, 1755, show:

"The officers to provide themselves with fuzils as the espontoon will be extremely inconvenient in the woods. The Sergeants are to be armed with firelocks and bayonets but to carry their swords.

"18 April, 1755. The men are to leave their shoulder belts, waist belts and hangers in store and are only to take 1 spare shirt, 1 spare pair of stockings and spare pair of shoes and 1 pair of brown marching gaiters."

Writing about this to Robert Napier, April 19th, 1755, Braddock says:

"As I shall find it often necessary to obliged the men to take with them 7 or 8 days' provisions, it being frequently impossible to supply them by the great distance from one magazine to another, and in order to enable them to carry any additional weight, I have lightened them as much as possible and have left in store their swords and the greatest part of their heavy accoutrements.

"Officers are to see that their men are provided as soon as possible with bladders or thin leather to put between the lining and crown of their hats to guard against the heat of the sun."

The Company Captains "are to have their arms in constant good order and every man to have a picker and brush, 2 good spare flints and 24 cartridges. The men are to apply to the train for paper and ball sufficient to complete every man with 24 rounds which are to be made up and distributed as soon as possible."

Another order says that "soldiers should be told that any man who upon the march by fastening his tent pole or by any other means incumbers his firelock shall be immediately and most severely punished."

An order dated Albany, May 5th, 1759, states that "officers will take fusil, no sashes but gorgets, either swords or hangers, as commanding officers of battalions shall direct. Regiments to take their colours into the field, the sergeants to carry firelocks, instead of halberts, with cartouche box and bayonet instead of sword, the soldiers no swords nor sword belt, if they can carry their bayonet securely without them; one Drummer per company; the remaining drummers to be put in the ranks; the Grenadiers to take their swords and caps into the field; no woman to be permitted to go with the regiments or to follow. The Royal Highland Regiment and the 77th Highlanders are excepted in the order of no swords. The C.O. of each of those regiments may do as he thinks best. The regiments to practise marching by files from the centre to the front, to halt and face outwards, match by files as before and form in battalion."

Amherst, Wolfe and other officers soon found that this new type of fighting in primeval forests necessitated some form of Light Infantry. To meet this need General Wolfe improvised the first light infantry.

Captain John Knox in his *Historical Journal of the Campaigns of* 1757, 1758, 1759 *and* 1760 quotes the orders for their formation as "a body of Light Infantry from the different corps to act as irregulars; the regiments that have been any time in America are to furnish such as have been most accustomed to the woods and are good marksmen and those from Europe are to furnish active marchers and men that are expert at firing ball and all in general must be alert spirited soldiers able to endure fatigue."

They were ordered "to exchange their heavy arms for those of the Artillery and the additional companies of Col. Fraser's Highlanders that are to remain in Nova Scotia to the number of 470." They were all to carry 70 rounds, the battalion companies having 50.

The force is stated to have been 550 which included 70 or 80 men of the Royal Americans and was placed under the command of Major Scott and entitled "The Light Infantry." They were given a special training similar to the Rangers for bush fighting, besides frequent target practice.

"The following order for the dress of the light infantry, as approved by his Excellency General Amhurst; Major-General Wolfe desires the same may be exactly conformed to by the light troops under his command; the sleeves of the coat are put on the waistcoat, and instead of coat sleeves he has two wings like the Grenadiers, but fuller; and a round slope reaching about half-way down his arm; which makes his coat of no encumberance to him, but can be slipt off with pleasure; he has no lace, but the lapels remain; besides the usual pockets, he has two not quite so high as his breast, made of leather, for ball and flints; and a flap of red cloth on the inside, which secures the ball from rolling out, if he should fall. His knapsack is carried very high between his shoulders, and is fastened with a strap of web over his shoulder, as the Indians carry their pack. His cartouch box hangs under his arm on the left side, slung with a leathern strap; and his horn under the other arm on the right, hanging by a narrower web than that used for his knapsack; his canteen down his back, under his knapsack, and covered with cloth, he has a rough case for his tomahawk with a button, and it hangs in a leathern sling down his side, like a hanger, between his coat and waistcoat, no bayonet, his leggins have leathern straps under his shoes, like spatterdashes; his hat is made into a cap with a flap and a button and with as much black cloth added as will come under his chin and keep him warm when he lies down; it hooks in the front, and is made like the old velvet caps in England."

These ear flaps seem to be like those shown for the proposed R.E. companies (Fig. 147, *see page* 196).

An authentic Account of the Reduction of Louisbourg published in London 1758 and quoted in *The History of the King's Royal Rifle Corps* gives the following curious description of these light troops, as being "chosen out of the most active and resolute men from all the battalions of regulars, dressed some in blue, some in green jackets and drawers for the easier brushing through woods with ruffs of black bearskin round their necks. The beard of their upper lips, some grown into whiskers, others not so but well smutted on that part, their arms a fusil cartouch box and powder horn."

The wearing of ruffs of bearskin is curious, but it is possible that they were to protect the ears and face in cold weather.

An order, "Camp at the Island of Orleans 1759," states that "The Light infantry of the army are to have their bayonets, as want of ammunition may sometimes be supplied with that weapon; and because no man should leave his post, under pretence that all his cartridges are fired, in most attacks by night it must be remembered that bayonets are preferable to fire."

Lord Howe of the 55th Foot went still further both in the actual training for forest fighting and in adopting suitable clothing. Later on, when we come to that part dealing with American Provincial troops, we will see how the skirts of the coats and the brims of the hats were cut off and instead of gaiters he clad his men in the leather leggings of the Rangers.

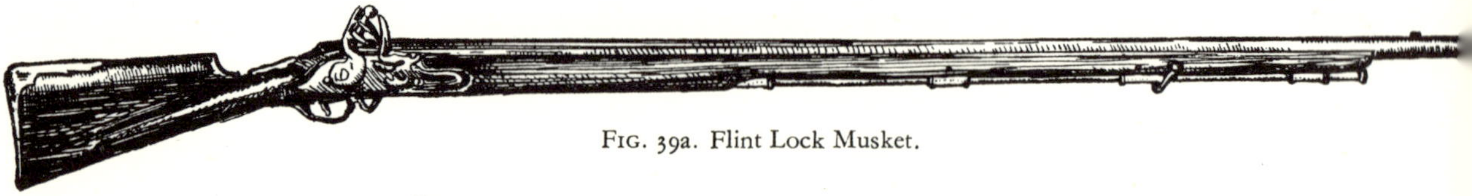

FIG. 39a. Flint Lock Musket.

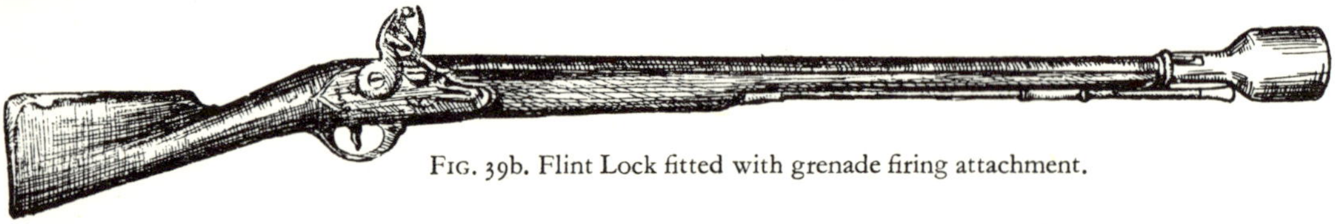

FIG. 39b. Flint Lock fitted with grenade firing attachment.

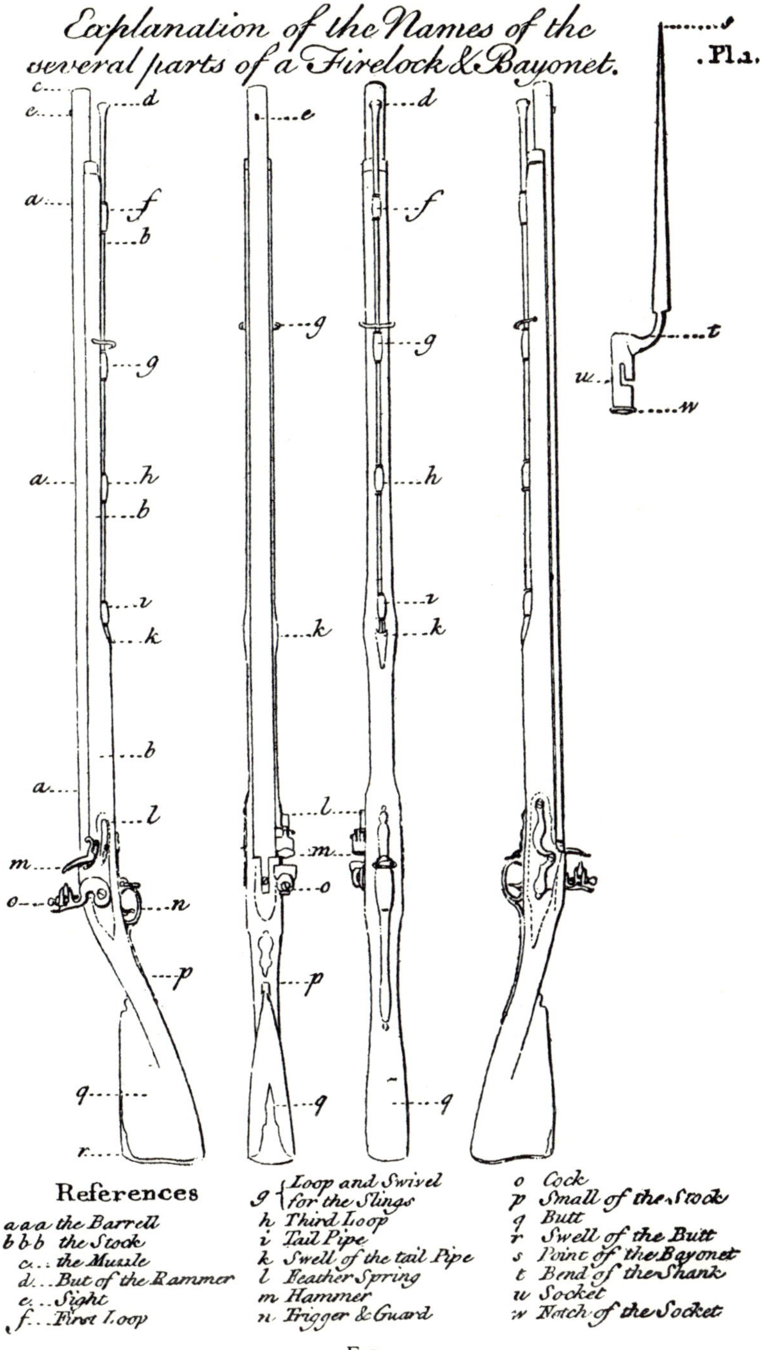
Explanation of the Names of the several parts of a Firelock & Bayonet.
.Pl.1.
References
a a a the Barrell
b b b the Stock
c... the Muzzle
d... But of the Rammer
e... Sight
f... First Loop
g {Loop and Swivel for the Slings
h Third Loop
i Tail Pipe
k Swell of the tail Pipe
l Feather Spring
m Hammer
n Trigger & Guard
o Cock
p Small of the Stock
q Butt
r Swell of the Butt
s Point of the Bayonet
t Bend of the Shank
u Socket
w Notch of the Socket

FIG. 40.

1740–1760

MILITARY music seems to have remained practically unchanged, that is, trumpets and kettle drums for regiments of Horse, hautbois and drums for Horse Grenadiers, Dragoons and Foot.

It is strange that such a delicate-toned instrument as the hautbois, or oboe of to-day, should have been so largely used for military music, but Farmer says that "it was then a much coarser instrument and was played with a reed almost as large as that used in a present-day bassoon."

The fife, which seems to have been laid aside for some time, is said to have been reintroduced into the army in 1748 after the Flanders campaign, both the Royal Artillery and the Guards claiming to be the first to readopt it. It rapidly became very popular with regiments, and drum and fife bands were formed in most regiments. This was partly the cause of retarding the formation of regimental bands. These do not seem to have made an appearance before 1750, and even then were composed of only a few musicians; and even in such regiments as the Guards they did not number more than six, two oboes or clarionets, 2 horns and 2 bassoons being their usual composition.

The clarionet, invented by a musician of Nuremberg in 1690, was first used by military bands, but when it was adopted in England, together with the bassoon and horn, Farmer is unable to say; he seems to suggest that it was in use in 1720–30. The French are said to have adopted it during the reign of Louis XV and acknowledged they copied it from the Germans. The Germans had for a long time been in advance of other countries in having bands attached to regiments and paid by the State.

In England the case was very different, as musicians were hired by the

FIG. 41. Bassoon.

DRUMS AND FIFES OF FOOT

1745 (after Hogarth) — 1751 (after Morier) — 4th King's Own 1745 — 18th or Royal Irish 1751

Colonel and paid by him and his officers. They were civilians and consequently not under any military discipline, and seem to have been in many cases a perpetual nuisance. Many were foreigners hired and brought from the Continent.

Fortescue says that in 1749 some effort was made to relieve Colonels in part of this expense "in hiring civilian bandsmen and in some regiments at any rate they were enlisted as soldiers and placed under military discipline."

A print *c.* 1750 (Fig. 43) shows the Foot Guards on the Horse Guards Parade, headed by a small band of musicians. Unfortunately the print is uncoloured and the figures too small to distinguish much detail of their uniform, but they appear to be dressed in coats similar to those worn by the rest of the regiment but without lapels and edged with lace down the front, button holes, pocket flaps and cuffs. The waistcoat and hat are also bound with lace. Judging by the various notices of the clothing for hautboys and drummers the uniform was probably a red coat faced blue; red waistcoats and blue breeches; and the lace no doubt was gold, in other words the Royal Livery, but not so elaborate as the state clothing. The musicians number eight and are marching in two ranks. In the first are two French horns, two bassoons, and the rear rank appears to be composed of hautboys or two hautboys and two clarionets.

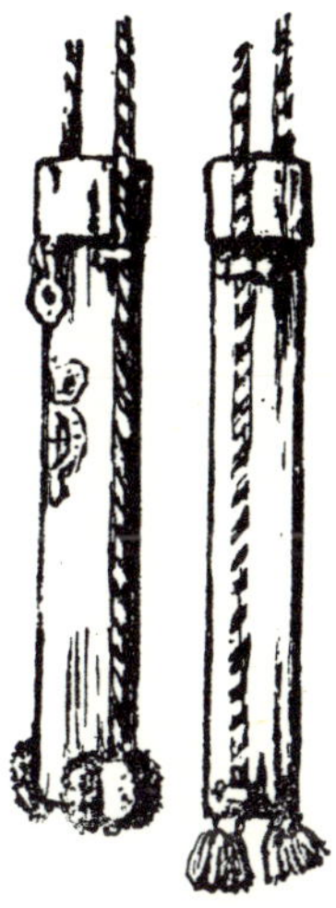

FIG. 42.
Fife Cases,
1745–51.

On glancing at the Inspection Returns of this period it will be seen that very few regiments as yet had bands of music other than drums and fifes.

The Warrant of 1751 states that drummers of all Royal regiments are allowed to wear the Royal livery, viz., red lined, faced and lapelled with blue and laced with a Royal lace. The drummers of all other regiments to have coats of the facing colour of the regiment, lined, faced and lapelled with red and laced as the Colonel shall think fit for distinction sake; the lace to be of the colour of that on the coats of the rank and file. The hautbois are not mentioned for the Infantry, but for the Dragoons they are to be dressed according to the same way as the drums, so we can only suppose that as a general rule musicians wore the dress of the

FIG. 43. Band of the Foot Guards, *c.* 1750.

drummers, but as they were largely paid out of the pockets of the Colonel and his officers, they no doubt had a free hand in dressing them as they liked, which, we will find later on, was often in a most elaborate manner.

The drummers' caps were similar to those of the Grenadiers, but instead of being stiffened they hung down at the back as previously in the form of a bag with a tassel at the end. The front was of the facing colour and embroidered with a trophy of drums and arms; a drum was also embroidered at the back.

Turning to the drums and fifes shown by Hogarth and Morier: Hogarth's drummers and fifers have been represented as having a small ostrich plume on their caps and on the fife case. On a careful examination of the original painting, however, these turn out to be the large worsted tassel on the cap and an ornament of red, white and blue worsted on the fife case, apparently as a decoration to finish off the case cords.

A comparison of Hogarth's and Morier's paintings leads one to suppose that the first artist has become involved with the ornamentation of the drummers' sleeves, as it seems most unlikely that a sleeve, even if made half red and half blue, would have been divided as he has shown, and that

Morier's is the real version of the sleeve, but with this exception they are most interesting as being the earliest detailed paintings of drums and fifes. It should be noted that although the caps painted by both artists agree in the details of trophies embroidered on the front, Hogarth shows the white horse on a blue little flap as it was worn previously to the red one painted by Morier.

Turning to Morier's drummers of the 4th or King's Own Regiment, the caps agree with details of trophies, but the lace seems blue with a white design, whereas the regiment's lace was white with a blue pattern (*see Plate*).

The drummer (back view) is thought to belong to the 18th or Royal Irish, as he is painted in the background of three Grenadiers and the only one of which could agree with the lacing and facings being the 18th. Here again the lace seems to be of blue with yellow spots and edging; the hanging sleeves are well shown.

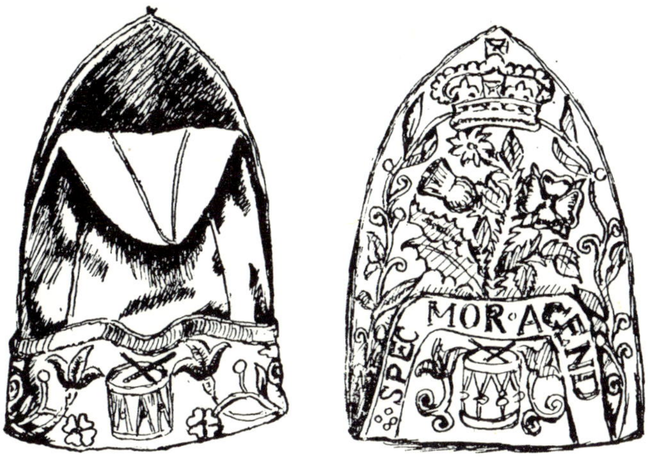

FIG. 44. Drum Major's Cap.

In the Royal United Services Museum is a Drum-Major's cap (Fig. 44). It is in a remarkable state of preservation. Both the front (13 inches high) and the little flap are a bright cherry red. In the centre are the rose, thistle and shamrock flanked by sprigs of laurel worked in natural colours. The little flap has a brown drum with red hoops, black cords and sticks. Round the edge is a white silk label bound with gold, bearing the motto *Spectemur Agendo* worked in black. The crown of the cap is yellow with the seams piped with red and the top falls over instead of being stiffened. The turn-up is red bound with gold lace and embroidered with a drum and foliage as the front. This cap is attributed to the 30th Regiment on account of the motto.

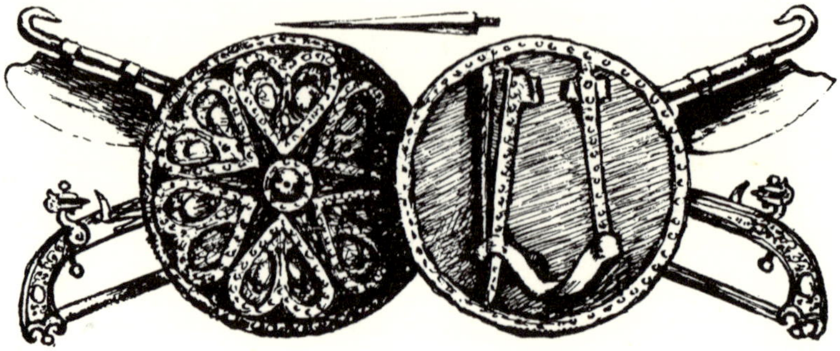

Fig. 45.

HIGHLAND REGIMENTS

c. 1714–1760

We have already seen that Independent companies of Highlanders had previously existed from 1667—in Dugald Mitchell's *History of the Highlands*, p. 595, the year 1624 is given as that in which the first independent companies were formed—and continued with short intervals up to 1717. They were then disbanded. They formed a species of military constabulary or gendarmerie for preserving order in the Highlands, especially after the fall of the Stuarts, and the consequent Jacobite unrest.

In 1725 they were re-raised, the officers and men coming principally from the Whig clans of Campbell, Grant, and Munro.

Owing to the fact that the Highland custom of carrying arms had been prohibited, many young men readily enlisted, so as to retain this privilege. Consequently, these companies were recruited from men of good family and character, and having some standing in their districts.

On May 15th, 1725, General George Wade issued the following order: "To Officers commanding Highland Companies;—That pursuant to their beating order they proceed forthwith to raise their non-commissioned officers and soldiers, and no man to be listed under size of 5 foot 6 inches. That the companies be complete by 15th June 1725.

"That Officers commanding companies take care to provide a plaid clothing and bonnet in the Highland dress for the non-commissioned officers and soldiers, belonging to their companies, the plaid of each company

FIG. 46. Private and Corporal of Highland Regiment, 1745.

to be as near as they can of the same sort or colour; that besides the plaid clothing, to be furnished every two years, each soldier is to receive from his captain a pair of brogues every six weeks, a pair of stockings every three months, a shirt and cravat every six months."

September 22nd, 1725, General Wade's orders to Lord Lovat or the officer commanding Highlanders:

"To provide them with such clothes and at such times as is mentioned in my former orders of May 15th, 1725. You are to send lists of the companies every four months to the officer commanding the troops quartered in the Highlands, viz. on 1st January, 1st May, and 1st September. The number of their badges to put before each man's name and when you have cause to change any of your men or fill up vacancies, you are to give the badge to the man who succeeds."

The question here arises, what were these badges?

The loyalist Highlanders in the "'45" wore a red cross in their bonnets, as a sign that they were in the Government service. From the above, however, it seems as if the badge mentioned was similar to the star given to the sheriffs of the old Western days of America, and also to those Indians who did the policing of the reservations.

It is considered that the Companies of 1725—1739 wore the red jacket and waistcoat, as certainly did the companies raised during the Rising of 1745–6. It should be remembered that these companies were on a much more organized and regular basis than those of the earlier period, besides having "colours" and being supplied by the Government with arms. The companies of 1678 had colours also. Clothing was to be supplied by the Captains of companies. The Companies of 1725—1739 assembled annually as a battalion.

Another piece of evidence advanced in support of this is the report of a

FIG. 47. Highland Soldier and his Wife, 1743–1750.

Court Martial in 1731 on two men involved in a brawl, John McNeall, of Lord Lovat's company, having discharged his pistol at James Simpson, of Captain Campbell's company.

McNeall is described as wearing a red coat and James Simpson a blue one. It seems that men of the companies were allowed to wear "coloured clothes," i.e. civilian clothes when off duty, as it is suggested Simpson was, his wife being a sutleress and the brawl having taken place in her hut or booth. The inference is that McNeall was in his red regimental coat and Simpson in civilian attire.

A History of the Highland Regiment, published in 1743, describes the arms and clothing of the Highlanders as follows: "To begin with their shoes, the Highlander wears a sort of thin pump or brogue, so light that it does not in the least impede his activity in running. In the next place, he wears broad garters under the knee and no breeches, but his plaid belted about his waist, which hangs exactly like the folds of the Roman garment, which we see in the equestrian statues; besides this, he wears a jacket with straight sleeves; as for his arms, they consist in the fuzil, a broad sword, a dirk or dagger, a Highland pistol all of steel, hung, on the other side of his belt opposite the dirk, and a target."

In 1725 a Drummer was ordered to be attached to each company, and the following story is related, by Burt, of a captain of one of the companies:

FIG. 48. Highland Soldiers, 1743–1750.

"He had received orders to add a drum to his bag pipe as a more military instrument. The pipe was to be retained 'because the Highland men would hardly be brought to march without it'; a dispute arose as to the post of honour 'which at length grew so exceeding hot' that the captain had to intervene, and decided in favour of the Drum, 'whereupon the piper remonstrated very warmly 'ods wuds (sic) Sir and shall a little rascal that beats upon a sheepskin tak the right hand of me that am a musician?' "

During the years 1745–1746 more Independent companies were raised. Major Mackay Scobie, in an article on the subject in the *Journal of Army Historical Research* (Spring, 1941), from which he has permitted me to quote, mentions that originally it was intended to raise twenty companies, but eventually only eighteen were formed. Their main duties were to consist in preventing the further recruitment of men, and levying of money and supplies for the "Pretender's service", "to live at discretion in the countries the rebels have left", and to march wherever and whenever deemed necessary "according to the orders they shall receive." The men were enlisted to serve until the final and total suppression of the Rising. Bounty or levy money was issued, as customary.

The allocation of the 18 companies was as follows:—

MacLeods (of Skye and Harris)	4	Grants 1
Mackenzies (Earl of Seaforth)	3	Munros 1
MacDonalds (of Sleat)	2	MacLeods (of Assynt) 1
Sutherlands	2	Rosses 1
Mackays (Lord Reay)	2	Inverness (Town) 1

From the correspondence between the Chief of the Gunns, who commanded the 1st Sutherland Company, and the Agent for the Companies, it is clear that a red jacket with yellow linings and facings, the usual bonnet, and plaid of a uniform tartan were worn. Owning to existing circumstances most of these Companies were not uniformed until after Culloden, wearing their own dress, with black cockade and a red cross (placed saltire-wise) in the bonnet, prior to that.

Here, as before, it is not clear if the same pattern tartan was worn by all the Companies, or each Company had its own sett; whichever the case, it is certain from the correspondence that special patterns were asked for from which to choose. Captain Gunn in one letter mentions "cloath and plaids for a hunder and ten men of the ordinary cloath the military gets, the Sergeants being a degree finer. I'll need 12 yards to each plaid. . . . The linings (of the coat) will be yellow with cloath cuffs."

Each Company was to be of a strength of 3 officers, 4 sergeants, 4 corporals and 100 private men, with a piper and drummer.

When on duty in the hills the heavy buckled shoes were laid aside, and the light *cuaran*, or heelless pump of untanned deerskin or cowhide, with the hair outside, was worn. These were fastened by leather criss-cross thongs over the instep and ankles.

Besides these 18 organised Independent Companies, there were "called out, armed, and subsisted, at various times" during the Rising, several hundreds of the Earls of Sutherland's and Seaforth's, Lord Reay's, Grant of Grant's, Munro of Culcairn's, Sir Alexander Macdonald's and Macleod's men, who were discharged when the occasion for such service ceased without any further payment. These wore their own dress, and arms, but were furnished in addition with muskets, bayonets and accoutrements by government.

The Argyll Militia (mentioned in Vol. I) was a complete enlisted Corps in itself, and was dressed in a uniform tartan, probably of the Black Watch or Military pattern. It, too, wore the red coat and waistcoat, and was issued by government with muskets, bayonets and equipment, besides their own arms (broadswords and dirks). As in the case of the Black Watch, Loudoun's Highlanders, and the Companies of 1745–6, side pistols seem to have been also an "issue", and not supplied by the men themselves.

We now come to the thorny and controversial subject of the tartans worn by the Independent companies and the Highland regiments subsequently raised. Much ink and no little heat has been expended on this subject, which in itself is a special and difficult study. I am therefore greatly indebted to Colonel M. M. Haldane and Major I. H. Mackay Scobie (Curator of the Scottish National Naval and Military Museum), who have made extensive researches into the subject, for their help and advice.

In the *Scots Magazine* of September, October and November, 1931, are a series of articles entitled "The Great Clan Tartan Myth," written by Colonel Haldane, which we would recommend for those wishing to go more fully into the matter.

Colonel Haldane has most kindly allowed me to use the information contained in these articles. As space does not permit going into all the arguments and evidence advanced, which has not been disputed or disproved, we will briefly note the main points necessary for our purpose.

Before starting it may be as well to explain that the word tartan originally meant a woollen material without any regard to its colour, and in fact continued to be so used by military tailors up to modern times. The Dress Regulations mention trousers of tartan which imply a plain mixture of almost black colour. The Gaelic word for tartan, meaning the chequered pattern as we now use it, was "breacan," but to avoid any confusion, the word tartan will be used in its modern and generally accepted sense.

The pattern or design of the tartan is called the sett, and is made up of a certain number of threads of each colour. The sett would vary in scale (i.e. size of checks) according to the thickness of the threads used, and also by reducing or increasing the numbers of threads proportionately. The general tone of the sett might also be effected by the shade of each colour being deeper or lighter without altering the sett.

FIG. 49. The Highland Regiment, 1743.
Officer. Sergeant. Private.

Colonel Haldane makes it clear that nowhere is there a mention by early contemporary writers, or foreigners travelling in Scotland, of a sett being the uniform or insignia of a certain clan.

One writer in 1700 states that even the local setts were not uniform and that "the plaids differed from each other as to the stripes and breadth of the colours," and "there is a great deal of ingenuity required in sorting the colours so as to be agreeable to the nicest fancy."

Another point one might bear in mind is that two branches of the same clan might be separated by a very considerable distance; and, taking into consideration the method of manufacture, and the lack of communication and the wellnigh impassable state of the country, one wonders how a uniform sett for the two branches was maintained.

FIG. 50. Highlanders, 1745.

Remembering the continuous clan feuds and the consequent state of more or less perpetual hostilities, a recognizable clan plaid would have been a positive danger to the wearer when outside his own territory.

The portraits of the seventeenth century show no sign of uniformity of tartan, and it is rare that any two members of a family are depicted wearing the same sett and in some cases the sitter is shown in articles of apparel made of two different patterns of tartan. In one family at least six different setts are shown and these differ entirely from those now known by the clan name.

The " '45" supplies no evidence that tartans were used as clan insignia. There is the case of the Macdonalds on the Government side meeting the Jacobite Macdonalds (who now wear different tartans), who were only distinguished by the cockades in their bonnets since the floral badge was the same.

The Campbell in the Government forces who had lost his bonnet with his badge and cockade and was in consequence in danger of being dispatched as one of the Jacobite Highlanders, is an incident showing that it was the badge or cockade which distinguished the sides, and not the tartan.

Relics of those tartans which were worn at Culloden or of the pre-1745 period bear no resemblance to any known modern tartan.

Morier's picture of Culloden, painted by order of the Duke of Cumberland and for which Highlanders taken prisoner at the battle were used as models, shows them wearing plaids, jackets and trews of different setts, while the tartans of no two men are identical (Fig. 50).

After the " '45," the wearing of Highland dress and arms was prescribed under heavy penalties, and this prescription was not repealed until 1782, although not enforced among those clans which had been loyal to the existing constitution.

The whole testimony of the clan tartan rests on the statements made by Stewart of Garth in his history of the Highlanders. Garth had been posted at the age of fifteen to the 42nd in 1787, forty-eight years after it was embodied, and wrote his book between the years of 1817 and 1822. He mentions that he had made no written notes but had to rely on his memory and those of others for his material, so that much of that which refers to the early period must have been at either third or fourth hand.

The final impetus to the clan tartan theory was given by the visit of George IV to Scotland in 1822. The ceremonial arrangements for the levée to be held were entrusted to Sir Walter Scott and Stewart of Garth, both romantically-minded men. They arranged that not only were the Highland gentlemen to be presented to appear in Highland dress, but that those of the same name or family were to wear the same "clan" tartan. Apparently there seems to have been considerable doubt in the minds of the chiefs themselves what tartan to wear, so Scott and Garth provided the details of the setts. Assistance was also afforded by the ingenious Sobieski Stuart brothers, *and* the tartan manufacturers.

To return to the Independent companies of 1725–39, Garth states that these wore the clan tartan of their captains, but from the foregoing this seems impossible; though some attempt at uniformity was made from the words: "The plaid of each company to be as near as they can of the same sort or colour."

That at first some sett was used, common to the locality in which the company was raised, is quite possible. A letter from Lord Lovat, dated 1727, states, however, that large quantities of tartan are available for any companies which required it. From this, and from the correspondence of Lord Lovat and other company captains between 1727–39, it seems to be fairly evident that a uniform tartan was adopted for all the companies.

One letter mentions that a surplus of tartan ordered for Lord Lovat's company is available for any other captains "who may want it." Sums of money are acknowledged as received from Lord Lovat for quantities of "tartan charged to Sir Duncan Campbell," and some of the captains of Highland companies. Another letter from Lord Lovat mentions "ye whole tartan for ye sex companies at 10 pence per yard."

In 1739, the Independent companies were formed into a regiment called "The Highland Regiment of Foot," and originally numbered the 43rd, later 42nd, or the Royal Highland Regiment, The Black Watch.

The uniform was a single-breasted, short red jacket without lapels, lined with buff, and a red waistcoat, both garments having white stitched buttonholes, which in the 1742 Clothing book are shown in pairs. The Van der Gucht and other early prints show no white edging to the buttonholes, only the 1742 Clothing Book. The buff facings are often described and depicted as a yellowish buff, but actual specimens of buff facing cloth found in Scotland (and also it may be added, an early pattern of those of "The Buffs") show a cream colour at times almost white in tone, which have made it difficult in some cases to decide if white or buff is intended for the facings of a regiment.

A flat blue bonnet which in some engravings appears to have a plain rim, and in others a red and white design. This Major Mackay Scobie suggests was the red border or rim, having a white ribbon threaded through a vertical slit in the rim for adjusting the bonnet to the head, and was, as much as the *fesse chequy* or dice board in the Royal Stewart arms, the probable origin of the dicing, which soon became a feature of the military bonnet. The 1742 Clothing book shows a small white patch like a piece of lace with a red zigzag pattern, placed on the left side of the bonnet. One wonders if this was not in reality the red cross badge of the Independent companies, before-mentioned. The bonnet on this figure is also furnished with a very small red tourie or tuft, which does not appear in any of the other engravings of the period. The bonnet has a black cockade and no other ornament.

The hose or caddis, from an old Scots word meaning chequered or striped, were made of white cloth striped with red, those originally worn by the regiment having a narrow black line edging the red. This pattern of hose tartan was thereafter called "the 42nd or old pattern" by the makers. The hose were cut to shape of the leg and sewn with a seam at the back and fastened with scarlet garters at the top without any turn over, or only a very narrow one.

The belted plaid was of the dark blue, black and green sett which, from its general use by the successive Highland corps raised, came to be regarded as the Government or military tartan. Colonel Haldane suggests that it was adapted "from an old sett in common use all over the Highlands, since it is the basis of many tartans which only differ by additional stripes of brighter colours, and as such would be suitable to a regiment recruited throughout the Highlands." That it was invented because the Earl of Crawfurd, the first colonel of the corps, was a Lowlander and therefore had no tartan, does not seem to have any foundation, as it appears Crawfurd was

FIG. 51. The Highland Regt., 1742.

on the Continent at the time the regiment was embodied. It seems permissable, however, to suggest that it was the tartan which had been worn by the Highland companies and continued to be used when regimented.

The Belted plaid, or the Breacan-an-fheilidh, was in fact the plaid and kilt in one, was made up of 12 yards double width of tartan. The way the belted plaid was worn is explained in the history of the Gordons as follows:

"The belt was laid down and the plaid placed over it, the centre of the

FIG. 52. Highlanders in Belted Plaid.

plaid being over the belt. It was then neatly pleated across the belt but leaving a part of each end unpleated. The wearer then lay down, and the belt was fastened round the waist so that the lower half of the plaid formed the feile or kilt, of which the unpleated part became the apron and the upper half falling over the belt, formed the breacan or plaid, which was fastened on the left shoulder with a pin or skewer. Later for military purposes it was looped up to the button on the left shoulder strap, or it could be thrown round the shoulders as a cloak. By loosening the belt, the whole became a blanket or plaid. (Plaid being the Gaelic for a blanket)" (Fig. 52).

The Feilidh beag or Little Kilt was worn for undress. This consisted of lower or pleated part but made up with the needle, the present kilt.

FIG. 53. 42nd or Highland Regt., 1743.

A red over stripe, running through green squares, was added to the tartan, said to have been introduced by Lord John Murray when he became Colonel in 1745. This red-striped sett appears to have been worn by the whole regiment when in the little kilt, and by the grenadiers only when in the belted plaid. The drummers, at an early date, had both scarlet and yellow stripes.

In some of the engravings of this time, the kilt appears to be split or divided in front and the pattern tartan placed diamond-wise and not at right angles. These, of course, are errors of the artists, who seem to have been mostly foreigners and therefore unacquainted with the intricacies of Highland garb. This curious rendering by early artists was from the fact that the old belted plaid, which was pleated at least once in front, had not the plain flat appearance of today. The dirk also hanging right in front helped to press in the front of the kilt and give it the divided appearance.

The Pipers, and later the Band of Music, wore a red tartan similar to or the same as that now known as the Royal Stewart. From its adoption by the bands of several Highland regiments it became known as the "Music tartan."

In the series of *Mutineer* Prints, published in 1743, Piper MacDonnel is shown with pipes having only two drones and ornamented with a banner showing a red cross in a yellow field (Fig. 60).

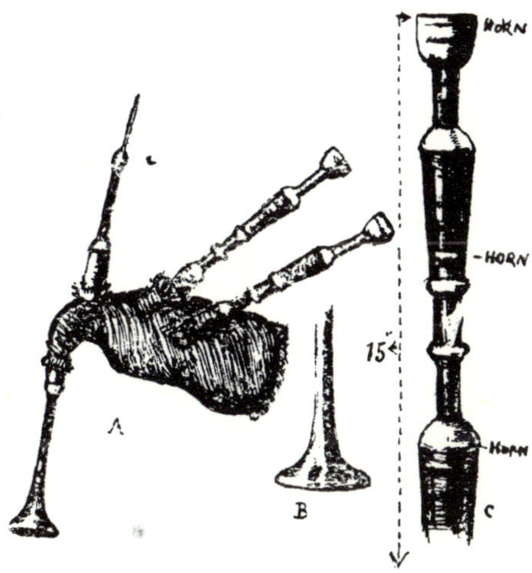

FIG. 54. Highland Pipes, *c.* 1745.

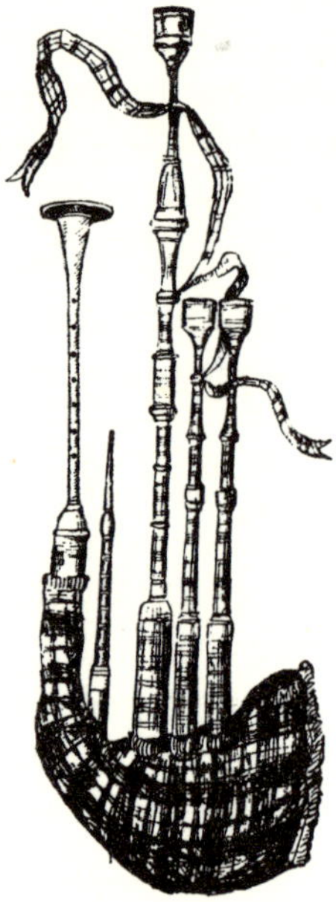

FIG. 55.
Highland Pipes, 1745.

The earlier types of pipe had usually only two drones, a third being adopted by North Highland pipers as early as the seventeenth century. The painting of William Cumming, Piper to the Laird of Grant, painted by R. Waitt, *c.* 1710, and the picture by Stoop, of the Demolition of the Mole at Tangiers, *c.* 1678, show both three drones and two drones. The bag-cover was of a plain colour, tartan not being generally adopted for this purpose until late in the eighteenth century, nor are the pipes usually shown decorated with tartan ribbons, which came in at a later date, superseding the coloured ribbons hitherto customary. (The early pipes were made of indigeneous woods, and had pear-shaped terminals.) (Fig. 54).

The chapter devoted to pipers and pipe music in the *History of the 93rd Highlanders* states that "the piper unlike the other musicians was also a fighting man. He commenced his duties at daybreak when after doffing his bonnet to the rising sun (a relic of the ancient sun worship), and repeating a short prayer, he roused the camp with the stirring notes of the Piob Mhor (Great Pipe)" (Fig. 55).

The Sporan, or purse, was made of otter, doe, seal, or deer skin, with the hair outside, but in some of the engravings it appears to be made of brown leather. It was usually furnished with a metal clasp-top (often ornamented) and usually had leather thongs knotted and frayed at the ends, in the form of tassels. An original purse of the Black Watch is in the National Naval and Military Museum, Edinburgh Castle.

The arms issued by the Government were the musket, bayonet and basket-hilted broadsword, the hilt being lacquered black. Besides these a dirk, and the claw-butted

FIG. 56. Highland Pistols.

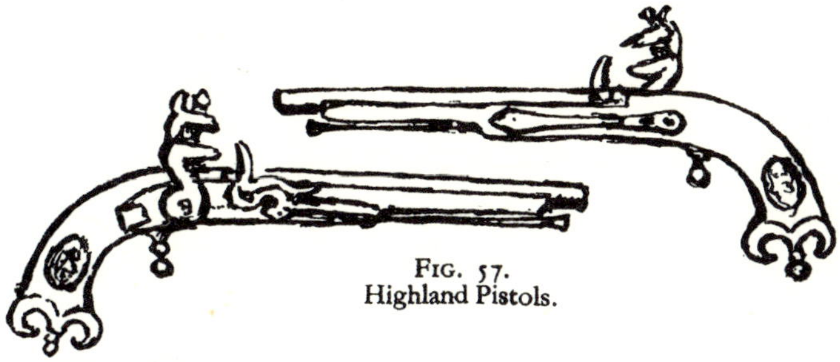

FIG. 57.
Highland Pistols.

Highland or side-pistol with iron or brass stock, and carried by a slide or hook on a narrow pistol belt. Targets (like the dirks) were permitted, and were furnished by the colonel or the men themselves. Grose relates that he remembers privates in the old Highland regiment carrying targets in Flanders, in 1745, which, though no part of their uniform, they were permitted to wear. These targets were studded with nails and were sometimes furnished with a spike in the central boss, which was detachable and carried when not in use in a small pocket or socket fixed to the back of the target. Targets were also used by the men of the Loudoun's Highlanders, and the Independent Companies of 1725–39, and 1745 (Fig. 45).

The dirk was carried in the left hand in such a way that the 16-inch blade covered and protruded beyond the forearm in a parallel position, so that when the arm was raised it could thus form a defence to parry blows besides being used for stabbing, much in the way shown in Morier's painting of the Highlanders (Fig. 60).

The whole system of fence with broadsword and dirk, and also with targe, or shield, was much like that of rapier and poignard, and buckler.

The Sergeants were armed with broadsword, dirk and halberd. It has been stated that they carried the Lochaber axe, but only halberds are mentioned in the lists of equipment and never the axe. This may have been carried by the original companies, as 69 Lochaber axes appear in the 1677 inventory of arms in Edinburgh Castle.

This weapon consisted of a broad axe blade and hook mounted on a 7 feet pole. It was not much used after 1715 (Fig. 45).

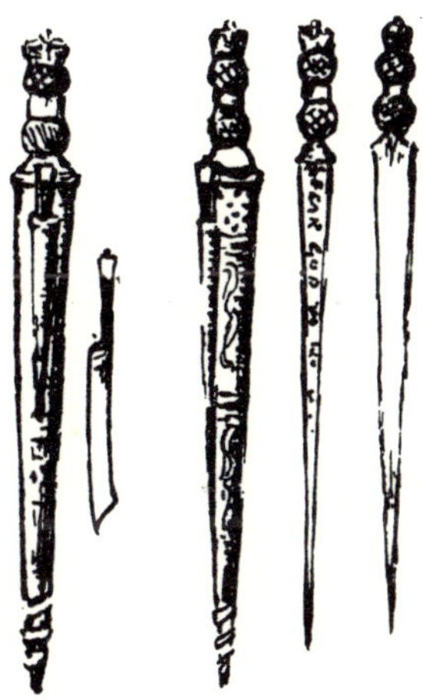

FIG. 57A. Dirk 1680.

FIG. 58. Highland Broad Sword, 1758.

The Corporal was armed like the privates, and was distinguished only by his white worsted shoulder knot, as in other regiments (Figs. 46 and 59).

The Pipers and Drummers were armed with broadsword (and possibly dirk) only.

The officers are represented carrying fuzils instead of the half-pike or spontoon, as was usual in the infantry at that time, the rest of their arms being broadswords, side pistols and dirks. The crimson sash was worn over the left shoulder and knotted on the right side, in much the same manner as worn to this day.

The officers' coats were laced with silver, and their sporans may have been more highly finished.

Although it has been stated that an eagle feather or tuft of bearskin was worn in the bonnets, there is nothing of the kind shown in the prints of this date.

In 1747, the Grenadiers of the Highland regiments were permitted, according to clothing orders, to wear fur caps instead of the cloth caps worn by the rest of the foot regiments. The other Highland regiment of this time was Loudoun's.

The title of "Black Watch," or "Am Freiceadan Dubh," it has often been stated, was given to the regiment on account of its dark tartan. It now appears, however, that there is no contemporary mention of its being generally so called, until towards the end of the eighteenth century and officially at an even much later date (1861).

There is evidence, however, that the title was occasionally applied to the Companies from which the regiment was formed (if not the earlier ones of 1667 to 1717). A letter dated April 7th, 1739, from a well-known officer in the companies, mentions that they were sometimes called "Black Watch." It seems quite likely the title was given in opprobrium by the Jacobite clans, owing to the force being composed of men of the Whig or Hanoverian party, and also to its duties of policing, or keeping watch, in the Highalnds and suppressing the practice of *blackmail.* It was unknown in the companies themselves.

It has already been pointed out that the regiment was recruited from totally different material to that of other corps, by the enlistment of men of good family. When first embodied it is related that it was no uncommon thing to see privates riding to the exercise ground followed by their servants carrying their arms and equipment. On one occasion, the King, wishing to inspect the dress of his new regiment, three privates remarkable for their figure and good looks were selected to appear at the Palace, where they also performed manual and platoon exercises before His Majesty, the Duke of Cumberland and General Wade, in the great gallery at St. James's. Each man received, it is said, a gratuity of a guinea which he gave to the porter at the palace gates, as he passed out, thinking that the King had mistaken their character and condition in their own country.

FIG. 59. Corporal of 43rd or Highland Regt., 1745.

In the night of February 8th–9th, 1749, five British transports with troops returning from Flanders were wrecked off the coast of France. Although the ships were entirely lost, the crews and passengers were not only saved but seem to have landed in complete uniform and equipment. Some of the survivors were sent to Bayeux and some to Caen, where they were lodged by the good people of those towns, three per bed in the inns and public buildings.

An eyewitness in a letter describes these soldiers as follows:

"These troops marched to fifres and bag pipes . . . the regiment of Lord John Murray was composed of 204 Highlanders having stockings of red and white cheques like harlequins, which stockings reached to two inches below the knee. The rest of the knee and three inches above, remaining uncovered."

"From the waist hung the kilt," and in the words of the writer, "Il y est

FIG. 60. Piper, 42nd or Highland Regt.

tres véridique qu'ils sont sans culotte dessous." "They wear a very short red jacket with white buttonholes, and a kind of cloak (plaid), very thin, of small squares of blue and green which covers half the body. The remainder is worn as a sash or scarf under the left arm and held up by the shoulder belt; on their heads they wear a small round very flat bonnet." In another place the bonnet is described as being made of blue cloth.

In the register at the Hôtel de Ville of Caen, there is also a mention of these ship-wrecked soldiers, and the fur caps of the Grenadiers are described as being large caps, ornamented or edged with fur, in the middle of which a cypher was embroidered in gold, which formed the letters G R interlaced.

FIG. 61. Field Officer of a Highland Regiment.

In the text it reads as if these fur caps were worn by Braggs's or 28th Regiment, but as we have seen that the Highland Regiments were the only ones permitted to wear fur instead of the usual cloth caps, it seems as if the writer had confused the two Regiments.

For pictorial evidence of the early Highland regiments, there is, first, the figure of the regiment in "The 1742 Clothing Book" (Fig. 51); second, the series of engravings by the two brothers Van de Gucht, who came over from Holland and were in London for some time. The date of their engravings is about

1743 (Figs. 46, 49, 52). In the same year Bowles published a set of "The Mutineers" (Figs. 55, 59, 60); a Private and a Piper by B. Gickman appeared, also dated 1743; a set by Müller, published by Bowles 1743–40. The last were "pirated" from the Van de Guchts. A series of a Highlander performing the manual exercises, published in 1743; a print entitled "The Hero of Fontenoy, 1745," showing a Highlander with his left arm cut off, leaving a gory stump and gripping a broadsword in his right hand, published 1745. Finally, are two of some set; one an officer and a sergeant, and the other a mounted officer and an ensign with regimental colour, signed J. C. Schmidhamer (Figs. 61, 66) and another print by him of various soldiers in pairs. The date of these is unknown beyond the fact that they appear to be between years 1743 and *c.* 1760, and were printed in Germany (Figs. 47, 48). It is supposed that if of these dates they represent the 42nd Loudoun's or Keith's and Campbell's Highlanders. Some of the figures in the latter series seem again to be inspired from the Van de Guchts. There is a very strong similarity in all the prints and no difference in the style of uniforms and equipment shown, so that they might be intended to represent any of the four Highland regiments mentioned. Three coloured prints (a soldier with his wife and child, a piper, and a drummer), appear in a series of European costumes, by Engelbrecht. These, from their German titles and description, represent Keith's and Campbell's Highlanders on the Rhine, 1759–63. The uniform is much the same as in above engravings; although the bonnet is more cocked (set up), with a red band, but has the large black cockade or bow, with no feathers or bearskin, however.

Fig. 62. Piper and Drummer of a Highland Regt.

After this, is the very important painting by Morier (1751) of a Grenadier of the 42nd. He wears the fur cap already mentioned; it has a red front decorated with a white ornamental edge and the Crown and G R. This front looks as if it might possibly be of painted or enamelled metal of a later date, but it is quite possibly of red cloth, embroidered in white like that worn by the other grenadiers of the line regiments. The jacket is single-breasted, with laced loops down the front, the lace having two red lines or stripes in it.

The rather large turn-down buff collar, which was not worn by other regiments at this time, is also edged with regimental lace, as well as the cuffs and slash. The shoulder strap is plain red. The waistcoat is red and edged with regimental lace.

FIG. 63. Grenadier, 42nd or Highland Regt.

The belted plaid is the usual dark government or military sett, with the addition of a red over stripe. The sporan is of plain leather with thongs and tassels of the same. The equipment and pouch is of black leather. The hose are the red and white pattern but show the diamond shape of check.

In recognition of His Majesty's satisfaction and approbation of the extraordinary courage, loyalty and exemplary conduct of the Highland Regiment, it was given the title of the 42nd, or "Royal Highland Regiment of Foot", by a special warrant dated July 22nd, 1758. The facings in consequence were changed to blue. This order arrived after the battle of Ticonderoga in Canada, July 8th, 1758, where the Regiment in its desperate attempts to force the French abattis, suffered casualties amounting to 314 killed and 333 wounded, out of a strength of 1,100.

While on service in Canada, the officers wore only narrow lace, or none at all, so as to be less conspicuous to the Indians in the forest fighting.

From the Order Books of the 42nd

Highlanders for the years 1759–60, when the regiment was in Canada, it appears that the plaid was used in camp, when off duty, for full dress parades, the men usually wearing the Feilidh Beag, or Little Kilt for undress and ordinary parades.

The officers wore plain red coats, white waistcoats and breeches and boots.

The Grenadiers and Pioneers were ordered to wear their caps at drills.

Before 1759 linen breeches for the officers, and short canvas drawers or breeches made from the tents, with blue leggings for the men were usually worn for field work and boat duty. On rare occasions the Feilidh Beag was worn with the blue leggings.*

In 1761 the Feilidh Beag with blue leggings is again mentioned.

In hot weather the waistcoat only was worn, both with the Feilidh Beag and breeches and leggings.

The bearskin tuft was ordered to be of the blackest fur procurable, and not to exceed 5 inches in length and to be fixed to the bonnet inclining towards the crown. The hair was to be worn queued or clubbed and powdered for full dress.

FIG. 64. Highland Bonnet.

Loudoun's Highlanders were raised in 1745 and took part in the defence of Bergen-op-Zoom, and were disbanded in 1748.

The Earl of Loudoun in his portrait by Allan Ramsay, painted in 1747, is said to be wearing the uniform of his regiment. The flat and round bonnet has a large black cockade; a red waistcoat laced with gold, brown purse with thonged tassels. The tartan is of blue, black and green with red over stripes. The coat has buff lapels which, however, do not appear to have been worn by Highland regiments until some years later. Millan's list, 1745, gives the regiment as wearing short coats, plaids, facings white, fur caps (for Grenadiers) (Fig. 65).

A contemporary painting of a Highland officer by Delacourt, of *c.* 1748, may represent an officer of this regiment, as the tartan is similar. The uniforms of the 42nd and Loudoun's are said to have been otherwise very much the same. He wears the flat blue bonnet with small bunch of feathers. The dirk is silver mounted, and the basket-hilted sword is carried in a black shoulder belt, with silver buckle, slide and heart-shaped tip. The hose are the military pattern, i.e. red and white and the facings a light buff.

* This may be compared with the wearing of "battle dress" instead of the Big Kilt in 1939.

FIG. 65. The Earl of Loudoun.

In 1759, the 87th Keith's and the 88th Campbell's Highlanders were raised and served in Germany during the Seven Years' War. Again there is little that is known of their uniform but they are supposed to have had buff or yellow facings, and a somewhat similar or the same tartan as the 42nd.

Keith had his portrait painted in Highland uniform, with coat of red, gold laced, faced green, but this was some years after the regiment was disbanded, and he was abroad at the time, and wrote home ordering Highland clothing to be made up and sent out to him. He is wearing an officer's coat and waistcoat of 1779. The tartan depicted is similar to that of the Black Watch: red and white hose.

The two regiments were disbanded in 1763.

Sir James Johnstone of Westerhall and other Highland gentlemen of Perth raised a regiment in 1760, known as the 101st or Johnstone's Highlanders. It at first sent drafts to Keith's and Campbell's Highlanders, but later served as a unit. Its facings were pale yellowish buff. Disbanded 1763.

The old 89th Highland Regiment was raised in 1759 by Major Staats Long Morris, husband of the Dowager Duchess of Gordon. It served in India and took part in the battle of Buxar. Its facings were light or pale yellow. and its tartan apparently the same as the 42nd. Disbanded 1765.

The old 77th or Montgomery Highlanders, was raised in 1757 by the Hon. Archibald Montgomery. It served in Canada and at Ticonderoga. The uniform was full Highland, the tartan being similar to that of the 42nd. The facings were at first red and were changed to green; the officers' coats and waistcoats being laced with silver. The regiment served in Canada, at Ticonderoga, against the Cherokee Indians 1760, at Dominique 1761, Martinique and Havanah 1762, and was disbanded 1763.

The old 78th or Fraser's Highlanders was raised by the Hon. Simon Fraser 1757, armed with musket, broadsword and dirk.

Stewart says "their bonnet was cocked on one side, slightly inclining

FIG. 66. Ensign, Highland Regt.

down to the right ear, and was furnished with two or more black feathers." The facings were light buff or white.

The painting of the death of Wolfe by Benjamin West, shows an officer of this regiment wearing a red jacket with buff facings, red and white hose, a flat blue bonnet with cockade, and a tartan plaid which appears to have a wide pattern, or sett of red, cream (or pale buff), blue or green. It must be acknowledged, however, that West made this painting some years after the event, and in fact the jacket has narrow buff lapels, which were not introduced until some years later. The regiment served in Canada, Newfoundland, and was present at Louisberg and Quebec.

It was proposed to change the uniform, as Highland dress was not considered suitable for the severe winters and summers of North America. Most officers and men "strongly opposed this, and were supported by Colonel Fraser." "Thanks to our gracious chief," said a veteran of the regiment, "we were allowed to wear the garb of our fathers, and in the course of six winters showed the Doctors that they did not understand our constitutions, for in the coldest winters our men were more healthy than those regiments who wore breeches."

The regiment was disbanded in 1763.

It will be noticed that in most cases where the tartan is mentioned, it appears to have been the military or government pattern, with or without coloured stripes. Later, in 1775–80, when many Highland regiments were raised, the badges, etc., used were sometimes those of the individual who raised them. There are also several cases mentioned in contemporary correspondence of Colonels, asking the manufacturers to make up samples of setts from which to choose the one to be worn by the new regiment.

These seem to be almost invariably of the Government tartan, plain or with one or more coloured stripes added. The books of the tartan manufacturers also show that the bulk of the tartan worn by regiments was the Government or military sett, with or without addition of over stripes. It was from the

FIG. 66a. Highland Dirk, *c.* 1745.

fact that a regiment was associated with a family or clan, that its tartan in the course of time became quite naturally to be considered that of the family which raised it.

In other words, instead of a regiment being dressed in a clan tartan, the clan adopted that worn by the regiment.

Not infrequently the drummers and pipers, at least, had a stripe in their tartan of the colour of the regimental facings.

It should be remembered that in many cases the noble or chief who raised a regiment was able to obtain a large proportion of his men from his own estates or those of his family, the remainder being recruited by his subaltern officers. In fact, the whole arrangements were very much on the lines of the old feudal and patriarchal systems, the result being that the relationship between officers and men was on a very different footing to that generally existing in other regiments of the army.

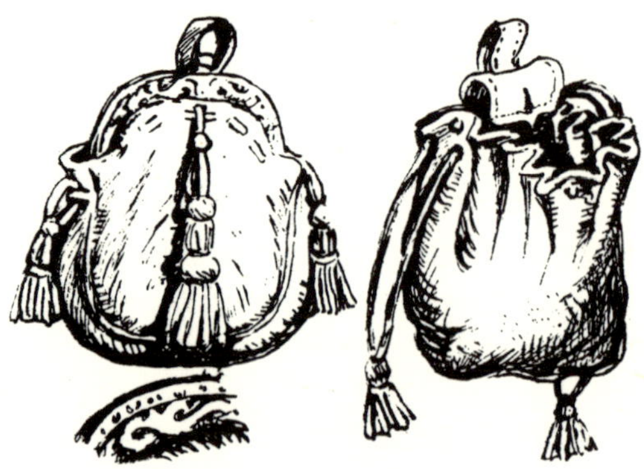

FIG. 66b. Early Types of Sporan.

SCOTS CORPS IN FOREIGN SERVICE

THE SCOTS DUTCH BRIGADE

1740–60

AFTER the rather scattered references to the dress of the Brigade mentioned in Vol. I we have little more until 1740, when we find that two of the regiments had yellow facings, and one, Mackay's, white. According to the contemporary notes of Leclerq in the Library at Mons, all the regiments had yellow facings. A sketch of a private of Mackay's in 1743, after papers in Ophemert Castle, show him still wearing the original white and so they seem to have remained for some time.

The "Ordre de Bataille" of the Dutch army gives the regiments as follows: Colyears 1 B^n, Lamy 1 B^n, Villegas 1 B^n Grenadiers. This regiment was composed entirely of grenadiers, during the campaigns of 1741 apd 1748.

When in 1749 the infantry was reorganised and all Battalions were composed of 2 grenadier and 8 Bn companies, all the companies of this regiment continued to wear the mitre-shaped cloth cap, until 1752.

The newly-formed regiments Dumlawig of 1 B^n was also a grenadier corps. The facings were green and all wore grenadier caps until disbanded in 1752.

In the reorganisation of 1752, when all single battalions were linked into 2 B^n regiments, we have the first official drawings of the uniforms as shown in the *Emplacement des troupes*. Then the series of a later date in an official book entitled *Uniformes qu'eurent les Troupes Hollandaises*, War Office, La Hague (Fig. 67).

The curious laces and tassels shown on the sleeves were a distinction of the Grenadiers and Drummers. The officers' uniforms were without silver or gold lace, which was strictly forbidden by the Stadholder in 1752. The hats only were ornamented with lace of either of these materials. A sash was worn over the right shoulder up to 1772; crimson for the regiments of Majoribanks and Stuart, and orange for Colyears. Gorgets of silver or gilt were worn when on duty. The company officers were armed with swords and spontoons, and those of the Grenadiers, light fusils (and swords).

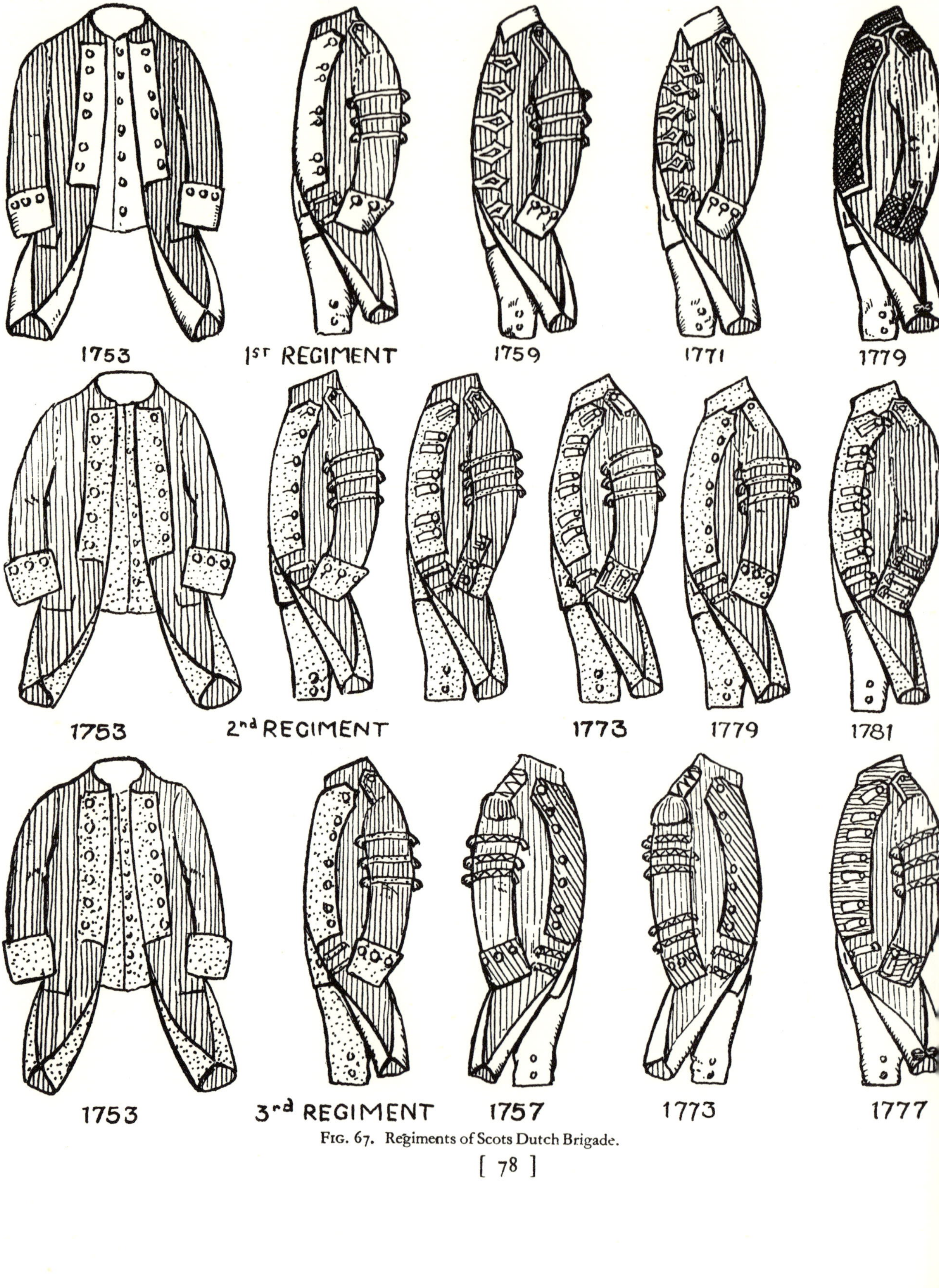

FIG. 67. Regiments of Scots Dutch Brigade.

A portrait of great interest is that of a Grenadier officer of a Scottish corps (an illustration of which we have been allowed to make by the courtesy of the owner, Mr. J. B. Wilkie, of Kirkintillock, Dumbartonshire) signed by the artist, and dated 1751. The front and little flap are a greenish tone, but this may have been blue originally. It is embroidered with the arms and supporters of Scotland, and is described by the Rev. Percy Sumner in his article in the *Journal of the Society for Army Historical Research* as follow: (Fig. 68):

Fig. 68. Grenadier Officer of a Scottish Regiment, 1751.

"Gold, a lion rampant within a double tressure flory-counter flory red; crest, on a crown, a red lion sitting front face, crowned gold, holding in his right paw a sword, and in his left a sceptre, both proper; supporters, two silver unicorns, horned, maned and hoofed gold, round the throat a gold coronet with crosses patty and fleur-de-lys, a gold chain affixed thereto between the forelegs and brought over the back. The supporters are accompanied by banners; the dexter being gold with red lion, and the sinister blue with a white saltire. At the top of the cap a silver scroll with the motto 'IN DEFENCE' and at the bottom a long silver scroll with 'NEMO ME IMPUNE LACESSIT'; in the centre of this scroll a blue circle bearing the badge of St. Andrew in silver. The little flap bears a silver grenade, with red flame inclined to the heraldic right; at each side are four thistles in red and green embroidery, arranged three above and one below.

"The cap is edged with a silver lace of 'fish scale' pattern.

"The coat is scarlet with yellow or buff cuffs, lapels and waistcoat of same colour and white breeches. The cuff is held up with a cord to the top button of the slash, in the same way as described in an earlier painting (Fig. 6). The cuffs, lapels and waistcoat are laced with silver. The sash is crimson, the gorget and sword hilt are gilt. Belts brown with gilt buckles.

"Unfortunately the regiment has not been identified up to the present.

"The presence in the background of Troopers in red coats and cloth grenadier caps, mounted on grey horses, at first led to the conclusion that it was an officer of the Scots Greys, but as far as is known the 'Greys' never bore the arms of Scotland on their caps and certainly did not have buff facings.

"It is also strange that the coat facings and cap front should be of different colours. One suggestion advanced is that this is a portrait of a Grenadier officer of the Scots Dutch Brigade."

The following notes are from lists of clothing supplied to the regiments:

1731—Colyears, red, yellow and green woolen twill, orange baize, pewter buttons, 126 grenadier caps.

1735—239 yds red and coloured cloth for the officers, 347 yds of red and coloured best woolen twill, 42 yds velvet, 256 yds green and orange woolen twill, 12 yds green cloth, 300 yds gold lace, 550 pairs of grey and red stockings for private soldiers, 550 privates' hats edged with false gold lace, 30 yds gold lace, 1192 doz brass coat buttons, 1375 doz brass waistcoat buttons.

1739—Red, yellow and green woolen twill, orange baize yellow baize, 25 yds green plush, 108 embroidered grenadier caps.

1743—60 yds green plush, red and yellow woolen twill, yellow baize, privates hats with false gold lace.

Majoribanks:

1755—244 yds red and white cloth for officers, 409 red and white cloth for sergts and Hautbois, 163 yds lace, gold and silver with knots and tassels, gilt buttons, brass and pewter buttons—188 pieces of red and white woolen twill, 56 pieces of red and white velvet, 516 hats edged white lace, 740 pairs of stockings, 126 yards broad lace for grenadiers and drummers, 103 cordonnets (?) 324 white tassels.

Halkets:

1739—3150 yds red woolen twill, 2640 yellow ditto, 58 yds grey ditto, 160 yds crimson cloth, 140 yds yellow ditto, 400 yds yellow serge, 83 yds gold lace, 450 lb yellow camel hair, 160 doz grenadier loops, 10 banderols (probably drum belts) 130 yds broad lace, 110 narrow lace.

1743—60 doz serge loops (for coats), 24 yds gold lace.

SCOTTISH REGIMENTS IN THE SERVICE OF FRANCE

WITHOUT going fully into the subject of the history of the Scottish corps, which served with the French armies in consequence of the ancient alliance between the two countries dating from the 14th century, the following may be briefly noted.

FIG. 69.
Archer Ecossais.

The Compagnie de Gendarmes Ecossais originated in the Gens d'Armes, or men-at-arms in full armour, who came to France during the endless wars of the 14th and 15th centuries. Later the Compagnie was further recruited from Jacobite nobles.

The uniform was scarlet with silver lace and, like the Scots Greys, they were mounted on grey horses.

The Archers Ecossais were formed by Charles VII, and a miniature painting by Jean Fouquet in the Musée de Chantilly is said to represent the corps in the time of Charles VII. This seems quite probable seeing that the body of men-at-arms behind Charles are shown wearing a more or less uniform dress and besides those armed with glaives and guisarmes a number are equipped with bows, arrows and pavis. They are in full armour and their côtes d'arms are of crimson, white and green divided into three equal parts, the centre portion white with crimson on the right, green on the left, decorated with heavy embroidery of gold and silver. The skirts are in alternate panels of the same colours. The short sleeves are a deep violet purple with groups of gold spots. This may be intended to represent pourpoint, although the sleeves seem too full in shape for this kind of defence (Fig. 69).

The salade is ornamented with an orle from which issue three plumes corresponding to the colours of the rest of the livery.

These Gardes Ecossaises endured up to *c.* 1839, but for a great number of years previously its ranks had been filled by Frenchmen only.

Among the staunchest followers of Jeanne d'Arc was a body of Scottish knights, and one French historian relates that the priest who carried the crucifix on her campaigns was also a Scotsman.

We find that a Scottish contingent was sent to France in 1589, and are thus described as a force of 12,000 foot and 60 horse under Dowin, landed at Dieppe in 1589: "Armed and clothed in the antique fashion in mail shirts and steel helmets, dressed in cloth and black priests caps, and playing on bagpipes and hautbois when going into battle."

A regiment entitled "Les Gardes Ecossaiss" was formed in 1641.

The following notes, gathered from French writers, state that the regiment was raised by Sir Robert Moray, "for service in the French Army." It distinguished itself at Rocroy (1643) and Lens. In 1662 it was incorporated in the regiment Douglas, formerly Hebron (Hepburn) which, as already mentioned, became the Royal Scots (The Royal Regiment).

Fig 70. Colour of Gardes Ecossais, 1643.

The "Colour" shown, therefore, has a connection with the last-mentioned regiment as being one of the first under which its forbears fought. The colour is described and illustrated in French works, as being "The cross of St. Andrew in a blue field, with the arms of France and Navarre on one shield, surrounded by laurel branches (or a golden collar) with the dove of the Saint Esprit at the base."

There are several points which are somewhat vague, as it seems unlikely that the dove of the St. Esprit would be attached to laurel branches.

The reconstructed standards which support the cradle of Henri IV show a shield and laurel branches similar to those on the colour of the Gardes Ecossaises but without the dove, so that it seems that the design was either the collar and dove of the St. Esprit or simply the laurel branches. Another point is the curious way the Arms of Navarre are emblazoned, which is not how the golden chains in a red field are usually represented.

Between the years 1744 and 1746, three Scottish regiments of infantry were raised in the pay of France, viz: Royal Ecossais, in France 1744; Ogilvy and Albany in France 1745.

The first consisted of one battalion of 12 companies and one of grenadiers. The uniform was a blue coat with red turn-down collar, shoulder strap, cuffs and linings, 11 double buttonholes of white lace, and buttons of white metal down the front of the coat, and 3 on the pockets. A red Fleur de Lys on the skirt turn backs. The veste or waistcoat was red, with ten white buttons and loops down either side of the front. The hat lace was white, white breeches and white gaiters.

It is mentioned both in the Army lists and personal memoirs of the time that the Grenadier company wore Highland dress and the kilt in particular is remarked on.

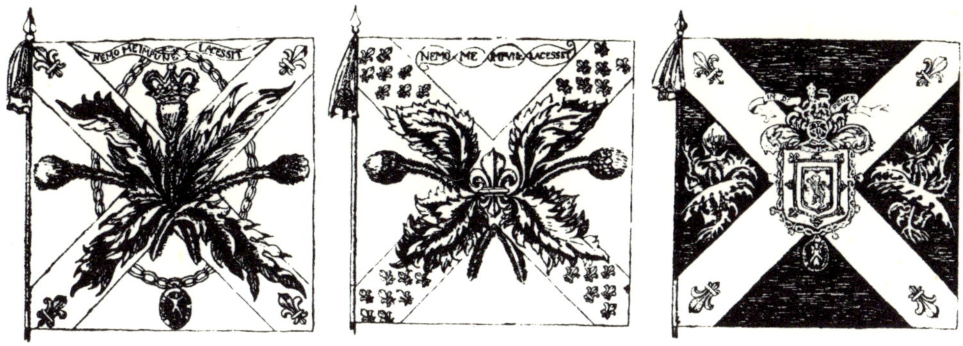

FIG. 71. Scottish Regiments in French Service.
Ogilvy. Royal Ecossais. Albany.

The regiment raised in 1745 by Lord Ogilvy wore very much the same uniform as the Royal Ecossais, but with the following differences:

All the buttonholes were without lace, and the coat pockets had five buttons. The skirt ornaments were the same. The Grenadier company also wore Highland garb.

These regiments were raised to be sent to Scotland in support of Prince Charles Edward, and after the failure of that rising remained in France and were disbanded in 1763, the men being incorporated into the Irish regiments of Bulkeley and Clare. Detachments served in the Rising of 1745.

A book of the period in the Library of the Ministère de la Guerre, Paris, contains large water-colour drawings of all the infantry colours of the French army, including these three Scottish corps.

Each is represented by a pair of colours: the 1st or Colonel's colour is white according to the French regulations, and the second blue and, like all French colours, ornamented with a cravate, or knot, of white silk with fringes, besides the tassels and cords, tied to the pole below the spearhead. Both colours are painted with the same design.

The Royal Ecossais has the cross of St. Andrew in white powdered with fleur-de-lys; in the centre a thistle and leaves painted in full realistic colours of dark green with blueish-grey high lights; in the centre of the cross is a large gold fleur-de-lys and on a white curiously chain-shaped scroll the motto "NEMO ME IMPUNE LACESSIT" in black.

Régiment d'Ogilvy. The St. Andrew's cross has in the centre a bunch of thistles and leaves painted in realistic colours, very much like the Royal Ecossais except the green is a lighter and more olive in tone. Over the top thistle is a curiously shaped crown in gold. Suspended from a white ribbon or scroll bearing the motto "NEMO ME IMPUNE LACESSIT" is a gold chain with a medallion attached on which is represented a nude figure of St. Andrew nailed to a cross, round which, on a blue edging is the same motto in gold letters. At the corner of each of the arms of the cross, over the whole face of the colour, is a gold fleur-de-lys.

Régiment d'Albanie. This is really a pair of colours of exceptional and rich design, especially so in the case of the blue colour. Over the whole surface is the cross of St. Andrew in white with a gold fleur-de-lys at each end of the arms. In the centre is a shield with the arms of Scotland in red and gold; above the shield is a steel helmet edged with gold and having green and gold mantling. The helmet is surmounted by the crown bearing the crown lion of England; a white banderol issues from either side of the crown bearing the motto "IN DEFENCE." The shield is surrounded by a gold chain having gold thistles at intervals; attached to the chain is a medallion similar to the one on Ogilvy's colour except St. Andrew is shown carrying his cross. In the spaces on either side of the cross is a thistle plant with blossom, leaves and roots painted in full realistic colours.

COLOURS OF IRISH REGIMENTS IN FRENCH SERVICE

OWING to the first Volume of this book having been divided into two these colours have lost their connection with the text as originally intended.

As mentioned in Vol. I, pp. 37 and 38, the Irish regiments in the Service of France were mainly descended from those regiments which came with James II, where they remained an independent army of that Monarch.

Even after their passing with the army of Louis XIV they still retained the red coat and Royal badges of the British army.

A B C

FIG. 71A. Colours of Irish Regiments In French Service.

Fitz James. The red cross of St. George bordered white, and a red saltire the whole in a green field. The motto is painted in gold.

Dorington. The red cross of St. George in a white field. In the centre of the cross is the Crown and Lion of Britain.

Bulkeley. A red cross bordered white. The 1st and 4th cantons are green and the 2nd and 3rd red. The crown, harp and motto in gold.

Clare. The same as Bulkeley, but the 1st and 4th cantons are yellow. It was into these last two regiments that the Scottish regiments of Ogilvy, Albany and Royal Ecossais were drafted after being disbanded.

Lally and Dillon are similar in design to Bulkeley except that Lally has the green and red cantons reversed and Dillon has black cantons.

INFANTRY COATS, FACINGS, LACE AND BUTTONHOLE LOOPS FROM CLOTHING BOOK, 1742

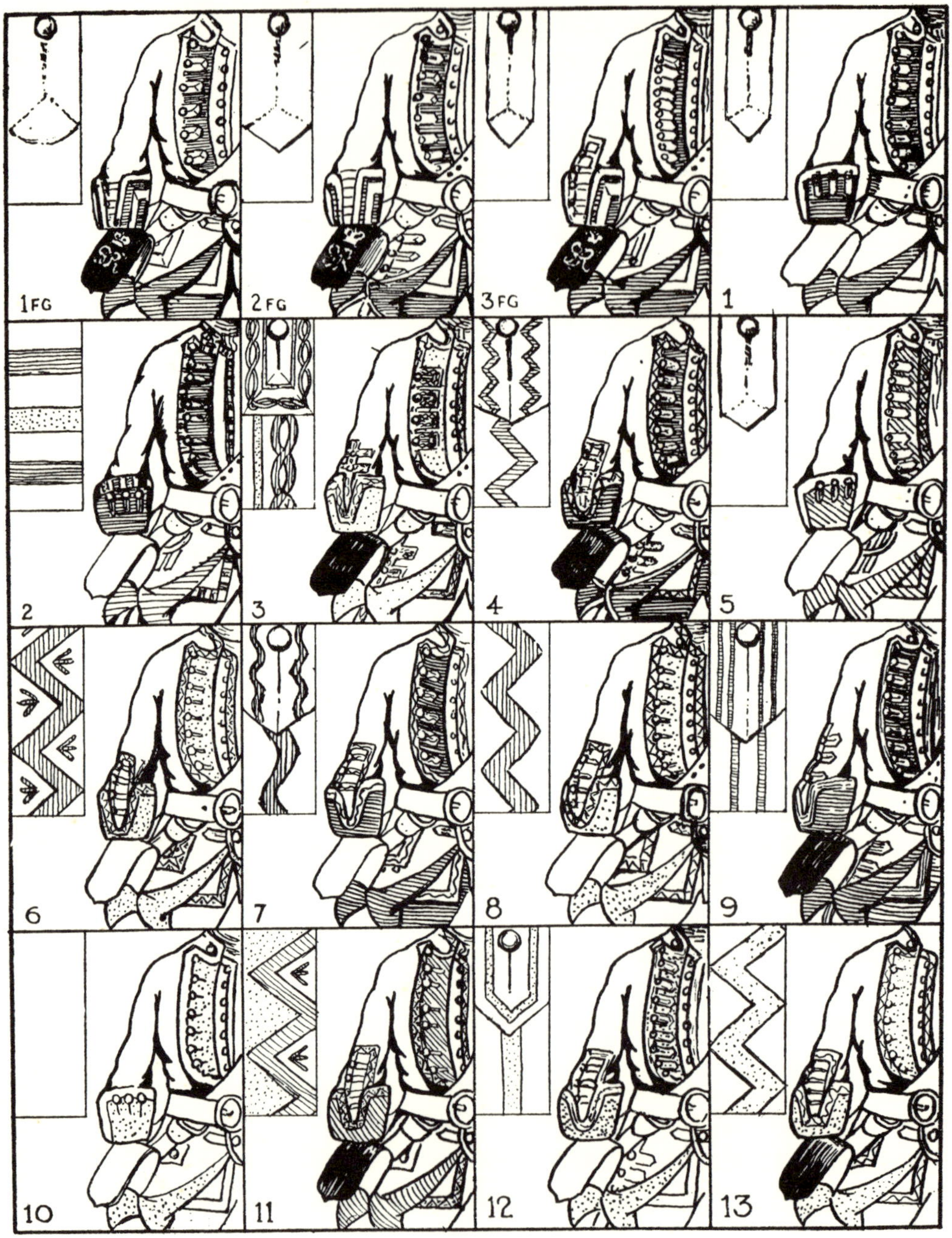

FIG. 72.

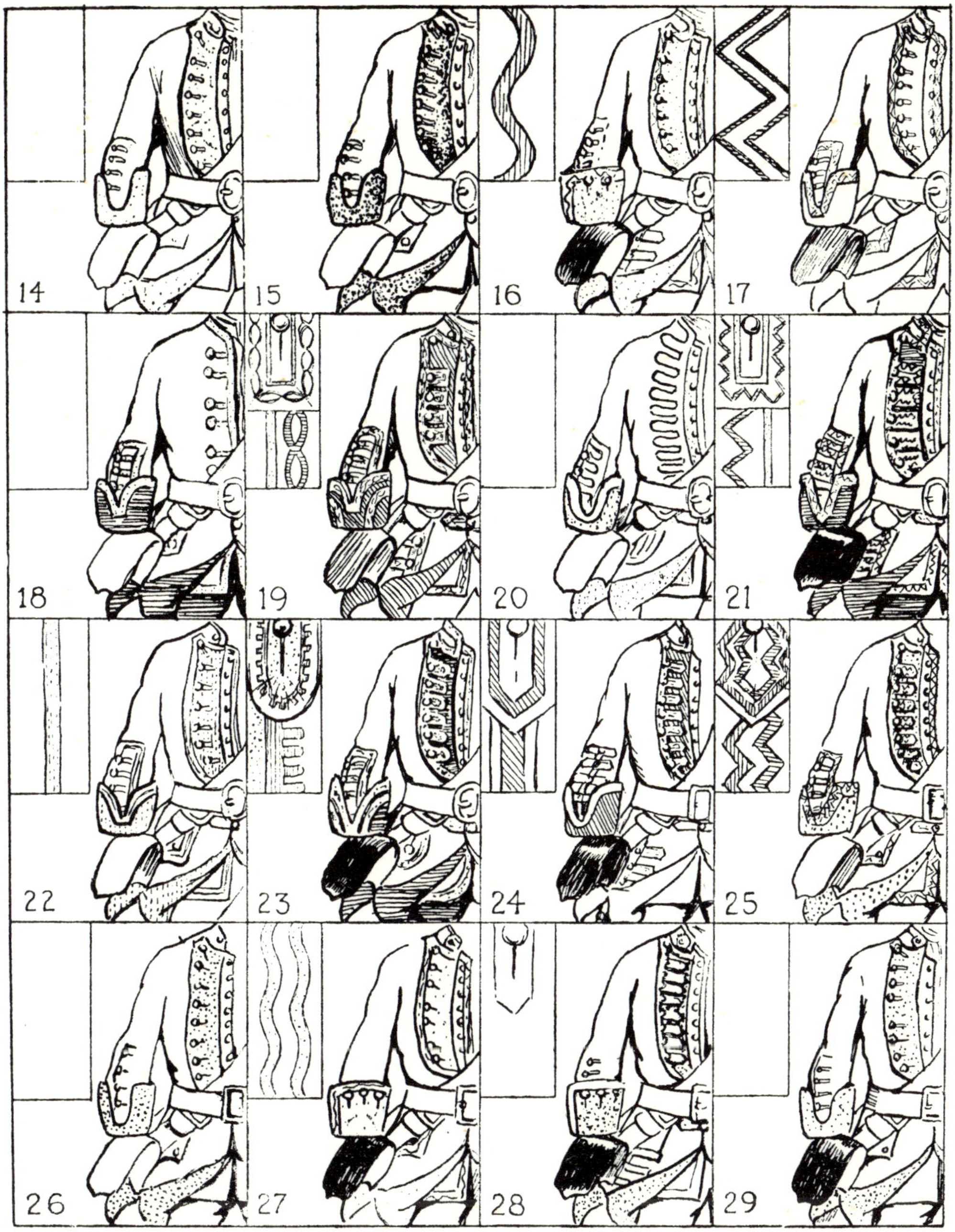

Fig. 73.

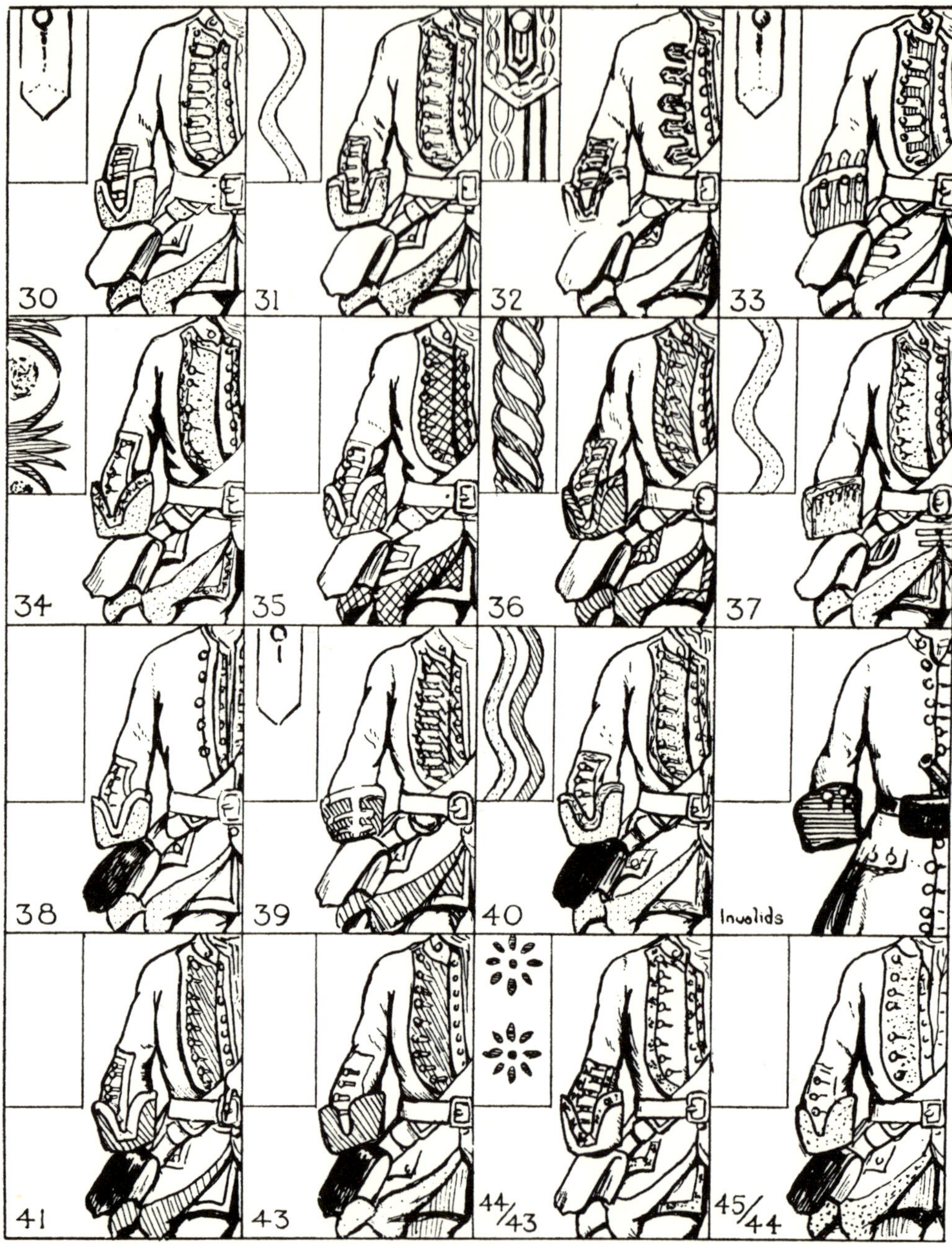

FIG. 74.

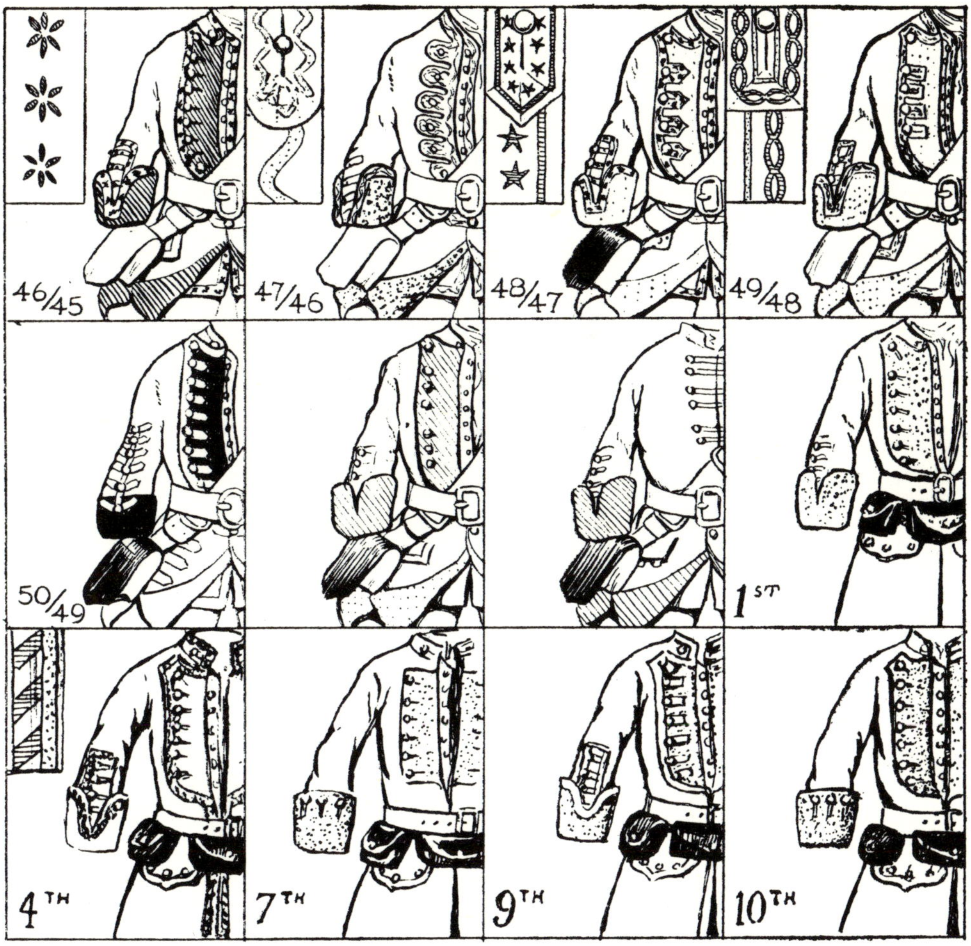

FIG. 75.

NOTE.—The two unnumbered figures are Independent Companies. The following numbered figures are the Marine Regiments.

INFANTRY COATS, LOOPS AND LACES *c.* 1750
FROM MORIER'S PAINTINGS

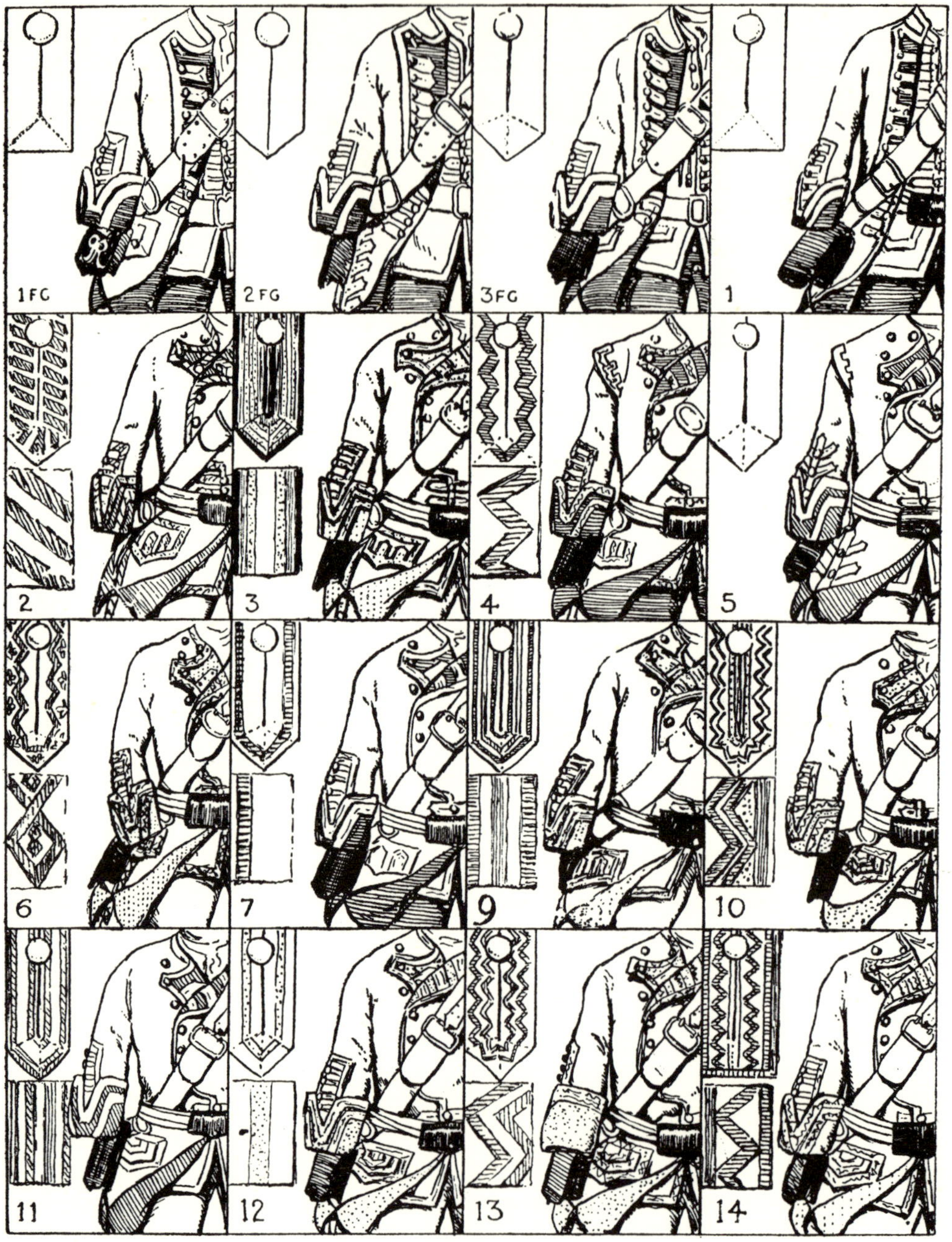

Fig. 76.

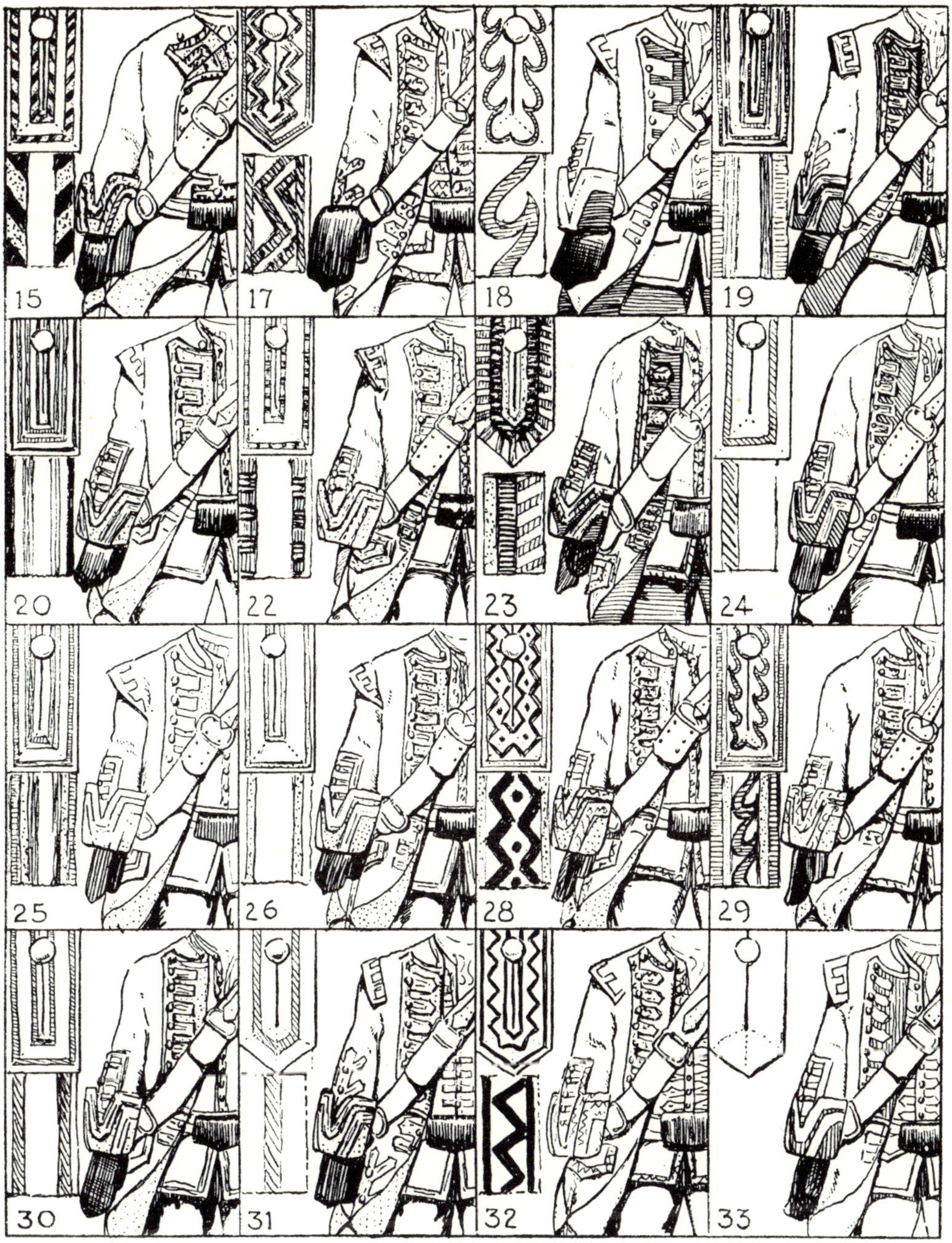

FIG. 77.

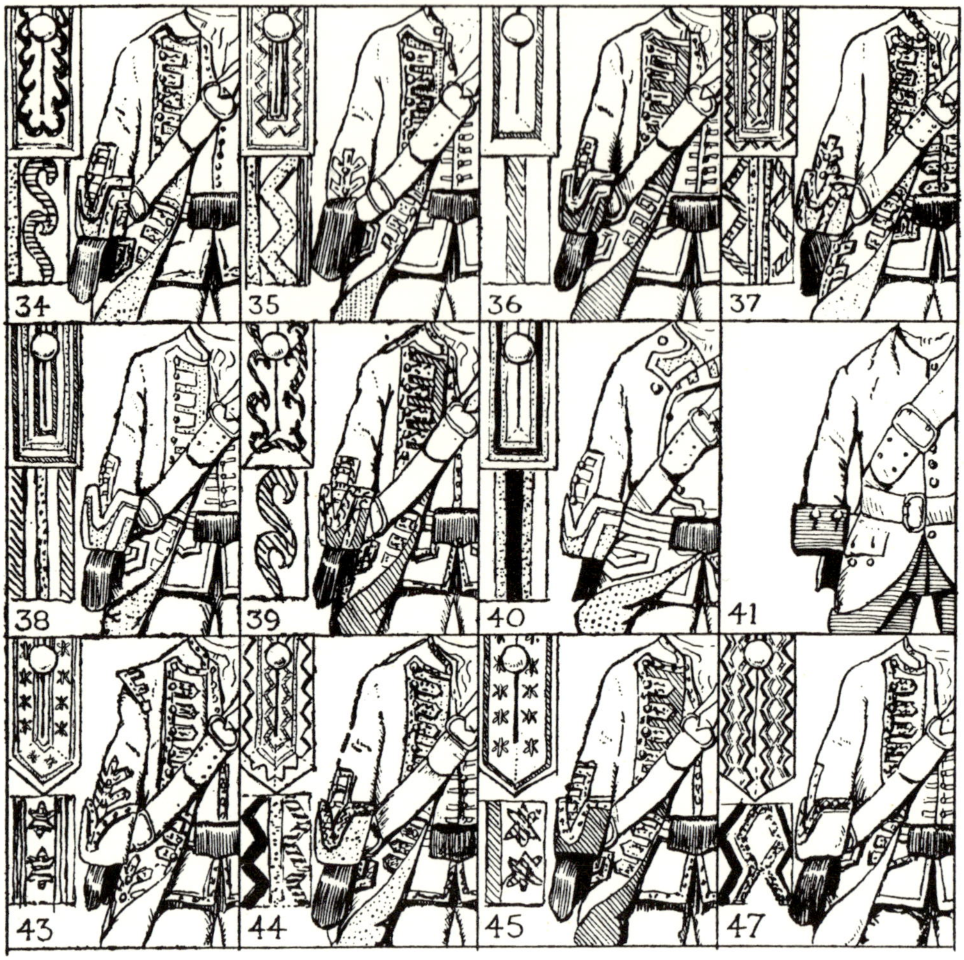

FIG. 78.

HERALDIC SYMBOLS FOR COLOURS

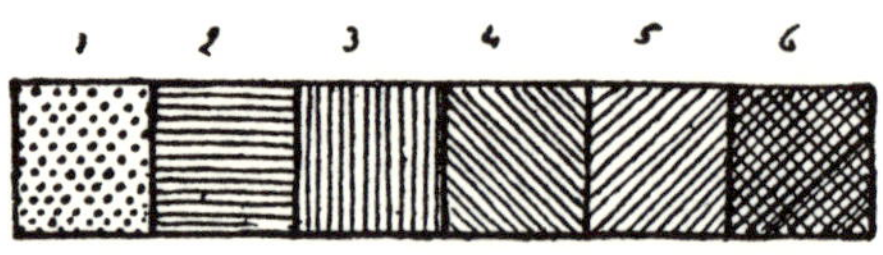

1. Or, gold or yellow.
2. Azure or blue.
5. Gules or red.
4. Vert or green.
5. Purpure or purple.
6. Orange.

APPENDIX: MILLAN'S LIST, 1749

The Facings of the Regiments of Foot

1st, or Royal Regt., blue
2nd, Tangier or Queen's, sea green
3rd, buff
4th, King's, blue
5th, pale green
6th, deep yellow
7th, Royal English Fuziliers, blue
8th, King's, yellow
9th, yellow
10th, yellow
11th, green
12th, yellow
13th, deep yellow
14th, buff
15th, deep yellow
16th, yellow
17th, light grey
18th, blue
19th, pale green
20th, pale yellow
21st, Royal Scotch Fuziliers, blue
22nd, reddish buff
23rd, Royal Welsh Fuziliers, blue
24th, green
25th, deep yellow
26th, Cameronians, yellow
27th, buff
28th, yellow
29th, yellow
30th, bright yellow
31st, buff
32nd, formed for Marines, white
33rd, red
34th, yellow
35th, orange
36th, green
37th, yellow
38th, yellow; brown linen linings for Leeward Is.
39th, pale green
40th, buff, formed of Independent Coys.
41st, blue from Invalids
42nd, Oglethorpe's, green, raised in Georgia and Carolina
43rd, Highlanders, buff, short coats and plaids; fur caps for grenadiers
44th, yellow, 1st Marines
45th, 2nd Marines, green
46th, 3rd Marines, yellow
47th, 4th Marines, white
48th, 5th Marines, pale yellow
49th, 6th Marines, green
50th, 7th Marines, white
51st, 8th Marines, yellow
52nd, 9th Marines, buff
53rd, 10th Marines, yellow
54th, white
55th, deep yellow
56th, deep green
57th, yellow
58th, white
59th, buff
60th, black, white linings
61st, green (Spotswood's) American

New Raised

61st, Richbell's raised 1742–43, white
62nd, Batereau, raised 1742, yellow
63rd, Trelawny's, green
64th, Loudon's, white, short coats, plaids, grenadier fur caps
65th, Shirley's, green
66th, Peperell's, green

On June 15th, 1758, the 2nd Battalions of fifteen regiments were ordered to be renumbered as separate units. In the following list, P.R.O. WO 7–26 they are all described as for the 61st except those with white, red, or black facings. For the sake of brevity a shorter formula has been used. The first numerals indicate the unit's new number, the second that of the regiment of which it had been the 2nd Battalion. All buttons are white.

61st (3rd). Buff facings and linings, white buttons, white lace with blue stripe. Drummers' coats buff.
2nd Colour buff with the union in upper canton. Rank on crimson in the centre.

62nd (4th). Facings, linings, drummers' coats and second Colour buff, white lace, 2 black and 2 buff stripes.

63rd (8th). Facings, drummers' coats and 2nd Colour very deep green, linings buff, white lace, green and white diagonal stripes. At Guadaloupe 1759 linen linings and thread or linen stockings.

64th (11th). Facings, drummers' coats black, linings white, white lace, 1 yellow stripe.
2nd Colour black, St. George's Cross throughout, union in the upper canton.

65th (12th). Facings, linings drummers' coat white, white lace with stripes of deep yellow, blue and red and a blue worm.
2nd Colour white with St. George's Cross throughout, union in upper canton, rank in centre on red.
1759 for Guadaloupe linen linings and stockings.

66th (19th). Facings, linings, drummers' coats and 2nd Colour yellowish green, white lace, 2 crimson stripes.

67th (20th). Facings, linings drummers' coats and 2nd Colour yellow, white lace, 1 yellowish green stripe.

68th (23rd). Facings, linings drummers' coats and 2nd Colour deep green, white lace, 2 yellow and 1 black stripe.

69th (24th). Facings drummers' coats and 2nd Colour willow green as 24th, linings white, white lace, 2 yellow and black stripes.

70th (31st). Facings drummers' coats and 2nd Colour deep grey, linings white, white lace, blue stripe.

71st (32nd). Facings, linings drummers' coats white, 2 black, 1 red stripe.

2nd Colour white with St. George's Cross, etc., as before.

72nd (33rd). Facings, drummers' coats red, linings white, lace white, 2 red and 1 black stripe.

2nd Colour white with St. George's Cross throughout, as before.

73rd (34th). Facings, linings, drummers' coats and 2nd Colour full green, white lace with white, blue and deep yellow stripes.

74th (36th). Facings, linings, drummers' coats and 2nd Colour deep green (like 36th), white lace with black and yellow figures; a footnote saying that this is a mistake, lace white, 2 red, 1 yellow stripes. While in Jamaica to have linen linings, ticken breeches and thread stockings.

75th (37th). Facings red, linings and drummers' coats buff; white lace, 2 green, 2 yellow stripes.

2nd Colour white with St. George's Cross throughout.

Further notes on other regiments from same records:

76th. Linen linings, ticken breeches, and stockings.

79th. For India, as 76th.

80th (Gage's). Facings and linings dark brown, short coats, black buttons, permitted to be without lace. Drummers' coats brown, No Colours.*

86th (1759). Facings deep orange lined brown linen, white buttons, lace white, 1 black stripe. Drummers' coats orange; thread stockings (at Goree).

*There has been much speculation about the interpretation of the above and what was the actual colour of the jackets of Gage's regiment. A contemporary statement, however, explains that these coats were red on one side and brown on the other and were so made that they could be worn with either side out, so that the regiment appeared in red, no doubt for dress parades, or in brown, this colour presumably being chosen as being indistinguishable for fighting in the forests of America.

REFERENCES FROM WARRANTS, BILLS, CLOTHING LISTS, INSPECTION RETURNS

3rd Foot, The Buffs.

1754–5. A band of music and two fifers.

Officers' uniform faced buff, bound and looped with silver. Inspection Returns.

4th, or King's Own Regiment.

1757. Red coats, lapelled and faced with blue, bound and looped with a white and blue lace, red waistcoats and blue breeches. Inspection Returns.

5th Foot.

1755. Red coats, lapelled, faced and lined with green, looped and bound with a mixed lace (white and green), red waistcoats and breeches.

Officers' coats lapelled, faced and lined pale green with a silver binding.

Regiment has fifes. Iron ramrods.

6th Foot.

1764. Silver mounted swords for sergeants.

7th or Royal Fuziliers.

1755. Officers' coats red faced and lapelled blue, gold lace binding, buff waistcoats and breeches.

Men, red coats faced, lapelled and lined blue, bound and looped with a mixed lace of white and blue, red waistcoats and breeches.

8th or King's Regiment.

1756. Drums and fifes. Red coats faced and lined blue, lace mixed white and yellow. For the Privates.

Officers' red faced blue and laced with gold.

9th Foot.

1733. "Col. Kane humbly besought the King that H.M. be pleased to grant him leave to change the facings of his regt's clothing from bright green to a light orange colour . . . H.M. hath thought fit to give his consent."

1755. Red faced and lapelled yellow, bound and looped with purple and white lace.

11th Foot.

1755. Drums and fifes. Men red coats faced, lapelled and lined green, looped with a mixed white, red and green lace.

1758. Officers, faced and lapelled green laced gold, red waistcoats and breeches.

12th Foot.

1755. Grenadiers had brass grenades on their pouches.

1756. Drums and fifes. Uniforms faced, lapelled and lined yellow, bound and looped with a mixed binding of white and yellow.

1758. Sergeants, gold laced hats. Men's with white worsted lace.

15th Foot.

1754. Silver lace and shoulder knots for officers.

1755. Facings, lapels and linings yellow, Lace, mixed yellow, black and white.

Drums and fifes.

18th Foot.

1755. Faced, lapelled and lined blue; blue breeches. A mixed white and blue lace.

20th Foot.

1755. Faced, lapelled and lined yellow. Lace a mixed black, white and red.

1757. Officers, silver lace.

1781. Drums and fifes.

23rd Foot.

"Deserter from Royal Welch Fuziliers, recruiting at Cambridge, in his regimentals—a plain red coat turned up with blue, yellow buttons, plain red waistcoat and blue breeches."—*Ipswich Journal*, 22 June 1745.

1757. On service sergeants carried halberts and officers fuzees. The latter had espontoons as well.

24th Foot.

1757. Facings willow green, linings white. Lace white and green.

Drums and fifes.

25th Foot.

1753. Gold lace and shoulder knots.

1756. Loops and binding white and yellow striped with red and blue.

27*th Foot.*

1733. Silver lace and buttons for officers.

28*th Foot.*

1733. Gold coloured buttons.

30*th Foot.*

1750. Silver lace and shoulder knots for officers.

1756. White and blue lace.

31*st Foot.*

1726. Silver lace for officers.

1730. Gold loops, shoulder knots, buttons and hat lace.

1733. Shoulder knots for sergeants.

33*rd Foot.*

1739. White linings. Silver shoulder knots, buttons, hat lace and loops for officers.

1757. Lace plain.

35*th Foot.*

1752. Silver lace, shoulder knots and buttons.

36*th Foot.*

1726. Officers, silver lace and buttons.

1756. Lace white and green.

37*th Foot.*

1734. Gold shoulder knots and buttons.

39*th Foot.*

1739. Silver lace.

52*nd Foot. Renumbered* 50*th*, 1757.

1756. Faced, lined and lapelled black, looped and bound white and black lace. Red waistcoat and breeches.

1758. Officers, red, faced and lapelled black velvet laced silver, white waistcoat and breeches.

Men's uniform lined white, white linen and black leather gaiters.

60*th*, *raised* 1755, *renumbered* 58*th* 1757.

Red, faced and lapelled with black lined buff. Marching gaiters only.

61*st*, *raised* 1755, *renumbered* 59*th*, 1757.

1756. Red, lapelled and faced Pompadour, lined red, yellow lace.

1759. Gilt buttons, broad and narrow gold lace; hats laced gold and gold loops for officers.

THE ROYAL CLOTHING WARRANT, 1751

GEORGE R.

Regulations for the Colours, Clothing, etc., of the Marching Regiments of Foot and for the uniform Clothing of the Cavalry, their Standards, Guidons, Banners, etc.

Our Will and Pleasure is That the following Regulations for the Colours, Clothing, etc., of Our Marching Regiments of Foot, and for the uniform Clothing of Our Cavalry, their Standards, Guidons, Banners, etc., be duly observed and put in execution, at such times as these particulars are, or shall be, furnished, viz., Regulation for the Colours, Clothing, etc., of the Marching Regiments of Foot.

No Colonel to put his Arms, Crest, Device, or Livery, on any part of the Appointments of the Regiment under his Command.

No part of the Clothing, or Ornaments of the Regiments to be altered after the following Regulations are put in execution but by Us, or Our Captain General's permission.

Colours.

The King's, or first Colour of every Regiment is to be the Great Union throughout.

The Second Colour, to be the colour of the Facing of the Regiment with the Union in the upper Canton; except those Regiments which are faced with Red or White, whose Second Colour is to be the Red Cross of St. George in a White Field, and the Union in the Upper Canton.

In the Centre of each Colour is to be painted, or embroidered in Gold Roman Characters, the Number of the Rank of the Regiment within a Wreath of Roses and Thistles, on the same Stalk, except those Regiments which are allowed to wear any Royal Devices, or ancient Badges, on whose Colours the Rank of the Regiment is to be painted towards the upper Corner.

The size of the Colours, and the length of the Pike, to be the same as those of the Royal Regiments of Foot Guards. The Cords and Tassels of all Colours to be crimson and gold mixed.

Drummers' Clothing.

The Drummers of all the Royal Regiments are allowed to wear the Royal Livery, vizt. Red, lined, faced, and lapelled on the breast with blue, and laced with a Royal lace: The Drummers of all the other Regiments are to be clothed with the Colour of the Facing of their Regiments, lined, faced, and lapelled on the Breast with Red, and laced in such manner as the Colonel shall think fit for distinction sake, the Lace, however, being of the Colours of that on the Soldiers' coats.

Grenadiers' Caps.

The front of the Grenadiers' Caps to be the same Colour as the facing of the Regiment, with the *King's Cypher* embroidered, and *Crown* over it; the little Flap to be Red, with the White Horse and Motto over it, "*Nec aspera terrent*"; the back part of the Cap to be Red; the turn-up to be the Colour of the Front, with the Number of the Regiment in the middle part behind.—The *Royal Regiments*, and the *Six Old Corps*, differ from the fore-going Rule as specified hereafter.

Drums.

The *Front* or forepart of the Drums to be painted with the Colour of the facing of the Regiment, with the *King's Cypher* and *Crown*, and the *Number* of the Regiment under it.

Bells of Arms.

The Bells of Arms to be painted in the same manner.

Camp Colours.

The Camp Colours to be Square, and of the Colour of the facing of the Regiment, with the Number of the Regiment upon them.

Devices and Badges of the Royal Regiments, and of the Six Old Corps.

1st Regiment, or The Royal Regiment.—In the *Centre* of their Colours, the *King's Cypher*, within the *Circle of St. Andrew* and *Crown* over it.—In the three corners of the *Second* Colour, the *Thistle* and *Crown*.—The *Distinction* of the Colours of the Second battalion is a flaming *Ray* of *Gold* descending from the upper corner of each Colour towards the centre.

On the Grenadier Caps, the same Device, as in the centre of the Colours; *White Horse* and the *King's Motto* over it, on the little Flap.

The Drums and Bells of Arms to have the same Device painted on them, with the Number or Rank of the Regiment under it.

2nd Regiment, or The Queen's Royal Regiment.—In the *centre* of each Colour the *Queen's Cypher* on a *Red* Ground, within the *Garter*, and *Crown* over it.—In the three corners of the Second Colour, the *Lamb*, being the ancient Badge of the Regiment.

On the Grenadier Caps, the *Queen's Cypher* and *Crown*, as in the Colours; *White Horse* and Motto "*Nec aspera terrent*" on the Flap.

The Drums and Bells of Arms to have the *Queen's Cypher* painted on them in the same manner, and the Rank of the Regiment underneath.

3rd Regiment, or The Buffs. In the *centre* of their Colours, the *Dragon*, being the ancient Badge, and the *Rose* and *Crown* in the Three corners of their Second Colour.

On the Grenadier Caps the *Dragon; White Horse* and *King's* Motto on the Flap.

The same Badge of the Dragon to be painted on their Drums and Bells of Arms, with the Rank of the Regiment underneath.

4th Regiment, or The King's Own Royal Regiment. In the *centre* of their Colours the *King's Cypher* on a *Red* ground within the *Garter*, and *Crown* over it: In the three corners of their Second Colour the *Lion* of *England*, being their ancient Badge.

On the Grenadier Caps the *King's Cypher*, as on the Colours, and *Crown* over it; *White Horse* and *Motto* on the Flap.

The Drums and Bells of Arms to have the *King's Cypher* painted on them, in the same manner, as the Rank of the Regiment underneath.

5th Regiment.

In the centre of their Colours, *St. George Killing the Dragon* being their ancient Badge and in the *three* Corners of their *Second* Colour the *Rose* and Crown.

On the Grenadier Caps, *St. George Killing the Dragon;* the *White Horse* and *Motto* "*Nec aspera terrent*" over it on the Flap.

The same Badge of *St. George and the Dragon* to be painted on their Drums, and Bells of Arms, with the *Rank* of the Regiment underneath.

6th Regiment.

In the *centre* of their Colours, the *Antelope*, being their ancient Badge, and in the three corners of their *Second* Colour, the *Rose* and *Crown*.

On the Grenadier Caps the *Antelope*, as in the Colours, White Horse and Motto on the Flap.

The same Badge of the *Antelope* to be painted on their Drums and Bells of Arms, with the Rank of the Regiment underneath.

7th Regiment, or The Royal Fusiliers.—In the *centre* of their Colours, the *Rose* within the *Garter*, and the *Crown* over it; the *White Horse* in the corners of the Second Colour.

On the Grenadier Caps, the *Rose* within the *Garter*, and *Crown*, as in the Colours; *White Horse* and *Motto* over it "*Nec aspera terrent*" on the Flap.

The same Device of the *Rose* within the *Garter* and *Crown* on their Drums and Bells of Arms, *Rank* of the Regiment underneath.

8th Regiment, or The King's Regiment. In the *centre* of their Colours the *White Horse* on a *Red* ground, within the *Garter* and Crown over it: In the three Corners of the Second Colour, the *King's Cypher* and *Crown*.

On the Grenadier Caps, the *White Horse* as on the Colours—the *White Horse* and Motto "*Nec aspera terrent*" on the Flap.

The same Device of the White Horse within the *Garter*, on the Drums and Bells of Arms; *Rank* of the Regiment underneath.

18th Regiment, or The Royal Irish. In the *centre* of their Colours, the *Harp* on a *Blue* field, and the Crown over it, and in the three Corners of their Second Colour, the *Lion of Nassau*. King William the Third's Arms.

On the Grenadier Caps the *Harp* and *Crown* as on the Colours, *White Horse* and *Motto* on the Flap.

The *Harp* and *Crown* to be painted in the same manner, on the Drums and Bells of Arms, with the *Rank* of the Regiment underneath.

21st Regiment, or The Royal North British Fusiliers. In the *centre* of their Colours, the *Thistle* within the Circle of *St. Andrew*, and *Crown* over it, and in the three corners of the Second Colour, the King's Cypher and Crown.

On the Grenadier Caps the *Thistle*, as on the Colours; *White Horse* and *Motto* over it "*Nec aspera terrent*" on the Flap.

On the Drums and Bells of Arms, the *Thistle* and *Crown* to be painted, as on the Colours, *Rank* of the Regiment underneath.

23rd Regiment, or The Royal Welch Fusiliers. In the centre of their Colours, the *Device* of the *Prince of Wales*, viz., three Feathers issuing out of the Prince's Coronet: In the three Corners of the Second Colour, the Badges of Edward the Black Prince, viz., *Rising Sun*, *Red Dragon*, and the three Feathers in the Coronet, Motto "*Ich Dien*."

On the Grenadier Caps the Feathers as in the Colours, White Horse and Motto "*Nec aspera terrent*" on the Flap.

The same Badge of the Three Feathers and Motto "*Ich Dien*" on the Drums and Bells of Arms; Rank of the Regiment underneath.

27th Regiment, or The Inniskilling Regiment.—Allowed to wear in the *centre* of their Colours a *Castle* with *three Turrets*, *St. George's Colours* flying in a *Blue* field, and the Name *Inniskilling* over it.

On the Grenadier Caps, the Castle and Name, as on the Colours: White Horse and King's Motto on the Flap.

The same Badge of the Castle and Name on the Drums and Bells of Arms, Rank of the Regiment underneath.

41st Regiment, or — The Invalids. In the *centre* of their Colours, the *Rose* and *Thistle*, on a *Red* ground, within the Garter, and Crown over it: In the three Corners of the Second Colour, the King's *Cypher* and *Crown*.

On the Grenadier Caps, Drums and Bells of Arms the same Device of the *Rose* and *Thistle* conjoined, within the Garter and Crown, as on the Colours.

Highland Regiment. — The Grenadier of the Highland Regiment are allowed to wear Bearskin-Fur Caps, with the *King's Cypher* and Crown over it, on a *Red* ground in the Turn-up, or Flap.

General View of the Facings of the Several Marching Regiments of Foot.

COLOUR OF THE FACINGS.	RANK AND TITLE OF THE REGIMENTS.	DISTINCTION IN THE SAME COLOUR.	NAMES OF THE PRESENT COLONELS.
Blue	1st or the Royal Regiment		Lt.-General St. Clair.
	4th or the King's Own Regt.		Colonel Rich.
	7th or the Royal Fusiliers		Colonel Mostyn.
	8th or the King's Regiment		Lt.-General Wolfe.
	18th or the Royal Irish		Colonel Folliot.
	21st or the Royal North British Fusiliers		Lt.-General Campbell
	23rd or the Royal Welch Fusiliers		Lt.-General Huske.
	41st or the Invalids		Colonel Wardour.
Green	2nd or the Queen's Royal Regiment	Sea Green	Major-General Fowke
	5th Regiment	Goslin Green	Lt.-General Irvine.
	11th Regiment	Full Green	Colonel Borland.
	19th Regiment	Yellowish Green	Colonel Lord George Beauclerk.
	24th Regiment (lined with White)	Willow Green	Colonel Earl of Ancram.
	36th Regiment	Grass Green	Colonel Lord Robert Manners.
	39th Regiment		Brigadier Richbell.
	45th Regiment	Deep Green	Colonel Warburton.
	49th Regiment	Full Green	Colonel Trelawny.
Buff	3rd Regiment or the Buffs		Colonel Howard.
	14th Regiment		Colonel Herbert.
	22nd Regiment	Pale Buff	Brigadier O'Farrell.
	27th or the Inniskilling Regt.		Lt.-General Blakeney.
	31st Regiment		Colonel Holmes.
	40th Regiment		Colonel Cornwallis.
	42nd Regiment		Colonel Lord Jno. Murray.
	48th Regiment		Colonel Earl of Horne

COLOUR OF THE FACINGS.	RANK AND TITLE OF THE REGIMENTS.	DISTINCTION IN THE SAME COLOUR.	NAMES OF THE PRESENT COLONELS.
White	17th Regiment	Greyish White	Lt.-General Wynyard.
	32nd Regiment		Colonel Leighton.
	43rd Regiment		Colonel Kennedy.
	47th Regiment		Colonel Lascelles.
Red	33rd Regiment (white lining)		Lt.-General Johnson.
Orange	35th Regiment		Lt.-General Otway.
Yellow	6th Regiment	Deep Yellow	Lt.-General Guise.
	9th Regiment		Colonel Waldegrave.
	10th Regiment	Bright Yellow	Colonel Pole.
	12th Regiment		Lt.-General Skelton.
	13th Regiment	Philemot Yellow	Lt.-General Pulteney.
	15th Regiment		Colonel Jorden.
	16th Regiment		Lt.-Genl. Handasyde.
	20th Regiment	Pale Yellow	Col. Lord Visct. Bury.
	25th Regiment	Deep Yellow	Col. Earl of Panmure.
	26th Regiment	Pale Yellow	Lt.-Genl. Anstruther.
	28th Regiment	Bright Yellow	Lt.-General Bragg.
	29th Regiment		Colonel Hopson.
	30th Regiment	Pale Yellow	Colonel Earl of London.
	34th Regiment	Bright Yellow	Colonel Conway.
	37th Regiment		Colonel Dejean.
	38th Regiment		Colonel Duroure.
	44th Regiment		Colonel Sir Peter Halket, Bt.
	46th Regiment		Colonel Murray.
Red with Blue Coats	Royal Regiment of Artillery		Colonel Belford.

Abstract of the Foregoing.

With Blue	. . . .	8 Regiments.
,, Green	. . . .	9 Regiments.
,, Buffs	. . . .	8 Regiments.
,, Yellow	. . .	18 Regiments.
,, White	. . . .	4 Regiments.
,, Red	. . . .	1 Regiment.
,, Orange	. . .	1 Regiment.
Blue with Red	. . .	1 Regiment.
In all	. . .	50 Regiments.

Regulation for the Uniform Clothing of the Cavalry—Their Standards, Guidons, Banners, Housings, and Holsters, Caps, Drums, Bells of Arms and Camp Colours.

Standards and Guidons

The Standards and Guidons of the Dragoon Guards, and the Standards of the Regiments of Horse, to be of Damask, embroidered and fringed with Gold, or Silver; the Guidons of the Regiments of Dragoons to be of Silk, the tassels and cords of the whole to be of crimson silk and gold mixed; the size of the Guidons and Standards, and the Length of the Lance to be the same as those of the Horse and Horse Grenadier Guards.

The King's, or First Standards, or Guidon of each Regiment to be crimson with the Rose and Thistle conjoined, and crown over them; in the centre His Majesty's Motto, *Dieu et mon Droit*, underneath; the White Horse in a compartment, in the first and fourth corner; and the rank of the Regiment, in gold or silver characters, on a ground of the same colour as the facing of the Regiment in a compartment in the second and third corners.

The Second and Third Standard, or Guidon of each Corps, to be of the Colour of the facing of the Regiment, with the Badge of the Regiment in the centre, or the Rank of the Regiment in Gold or Silver Roman Characters on a crimson ground, within a Wreath of Roses and Thistles on the same stalk, the Motto of the Regiment underneath; the White Horse on a Red ground to be in the first and fourth Compartments, and the Rose and Thistle conjoined upon a red Ground in the Second and Third compartments.

The Distinction of the Third Standard, or Guidon to be a figure 3 on a circular ground of Red, underneath the Motto.

Those Corps which have any particular Badge, are to carry it in the centre of their Second and Third Standard or Guidon, with the rank of the Regiment on a red ground, within a small wreath of Roses and Thistles, in the second and third corners.

Banners

The Banners of the Kettle Drums and Trumpets to be the colour of the facing of the regiment, with the badge of the Regiment, or its rank, in the centre of the banner of the Kettle Drums, as on the second Standard; the King's Cypher and Crown to be on the banners of the trumpets, with the rank of the regiment in figures underneath.

Drums

The Drums of the Dragoon Guards, and Dragoons to be of brass; the front, or forepart, to be painted with the colour of the facing of the regiment, upon which is to be the badge or rank of the regiment as in the second guidon.

Bells of Arms

The Bells of Arms to be painted in the same manner as the Drums.

Camp Colours

The camp Colours to be of the colour of the facing of the regiment, with the rank of the regiment in the centre—those of the Horse to be square, and those of the Dragoon Guards, or Dragoons, to be swallow-tailed.

Clothing of the Regiments; Distinction of the Serjeants and Corporals; Clothing of the Kettle Drummers, Trumpeters, Drummers, and Hautboys, Drummers' Caps

The Coats of the Dragoon Guards to be lapelled to the Waist with the colour of the regiment, and lined with the same colour; Slit Sleeves, turned up with the colour of the lapell.

The Coats of the Horse to be lapelled to the bottom with the colour of the regiment and lined with the same colour (except the fourth regiment of Horse, whose facings are black, and the lining buff colour), small square Cuffs of the Colour of the lapell.

The Coats of the Dragoons to be without lapells, double breasted, slit sleeves, turned up with the colour of the facings of the regiments, the lining of the same colour.

The whole to have long Pockets, the buttonholes to be of a very narrow yellow or white lace, as hereafter specified, and set on two, and two, or three and three, for distinction sake. The shoulder knots of the Dragoon Regiments to be of yellow or white worsted, and worn on the right shoulder. The waistcoats and breeches to be of the colour of the facings, except those of the Fourth Regiment of Horse which are Buff colour.

The Serjeants of the Dragoon Guards and Dragoons to be distinguished by a narrow gold or silver lace on the lapells, turn up of the sleeves and pockets, and to have gold or silver shoulder knots. The Corporals of Horse by a narrow gold or silver lace on the lapells, cuffs, pockets and shoulders straps; the Corporals of Dragoon Guards and Dragoons by a narrow silver or gold lace on the turn up of the sleeves, and shoulder strap, and to have yellow, or white silk shoulder knots.

The Kettle Drummers', Trumpeters', Drummers' and Hautboys' Coats to be of the colour of the facing of the regiment, lined and turned up with red (except the Royal Regiments which are allowed to wear the Royal Livery, viz., red, lined, and turned up with blue—blue waistcoats and breeches) and laced with the same coloured lace, as that on the Housings and Holster Caps, red Waistcoats and Breeches. The Drummers and Hautboys of the Dragoon Guards, and the Kettle Drummers and Trumpeters of the Horse to have long hanging sleeves fastened at the Waist.

The Caps of the Drummers to be such as those of the Infantry, with the Tassel hanging behind, the front to be of the colour of their facing with the particular badge of the regiment embroidered on it, or a trophy of guidons and drums, the little flap to be red, with the White Horse and Motto over it, *Nec aspera terrent;* the back part of the Cap to be red likewise; the turn up to be the colour of the front, and in the middle part of it behind, a Drum, and the rank of the regiment.

Hats and Caps of the Cavalry	The Hats to be laced with gold or silver lace, and to have black cockades. The Royal North British Dragoons, only, to wear caps instead of Hats, which Caps are to be of the same form as those of the House Grenadier Guards; the front Blue with the same badge as on the Second guidon of the Regiment; the flap red, with the White Horse and Motto over it, *Nec aspera terrent;* the back part to be red, and the turn up blue, with a Thistle embroidered, between the letters II.D. being the rank of the Regiment. The Watering or Forage Caps of the cavalry to be red, turned up with the colour of the facing, and the rank of the regiment on the little flap.
Cloaks	The Cloaks to be red, lined as the Coats, and the Buttons set on at top, in the same manner, upon frogs, or loops of the same colours as the lace on the Housings, the capes to be the colour of the facings.
Housings and Holster Caps	The Housings and Holster Caps to be of the Colour of the facing of the Regiment (except the First Regiment, or King's Dragoon Guards, and the Royal Dragoons, whose Housings are red, and the Fourth Regiment of Horse, whose Housings are buff colour), laced with one broad white or yellow worsted or mohair lace with a Stripe in the middle of one third of the whole breadth, as hereafter specified: the Rank of the Regiment to be embroidered on the Housings upon a red ground, within a wreath of Roses and Thistles, or the particular badge of the Regiment as on the Second Guidon or Standard: The King's Cypher, with the Crown over it, to be embroidered on the Holster Caps, and under the cypher the number or rank of the Regiment.
Uniform of the Officers, etc.	The Clothing or Uniform of the Officers, to be made up in the same manner as those of the Men, laced, lapelled, and turned up with the colour of the facing, and a narrow hold, or silver lace, or embroidery, to the binding and buttonholes, the buttons being set on in the same manner as on the Men's Coats; the Waistcoats and Breeches being likewise of the same colour as those of the Men. The Housings and Holster Caps of the Officers to be of the colour of the facing of the Regiment, laced with one gold or silver lace, and a Stripe of velvet in the middle of the colour of that on the Men's. The Standard belts to be the colour of the facing of the Regiment and laced as the housings. Their Sashes to be of crimson silk, and worn over the left Shoulder. Their Sword Knots to be crimson and gold in stripes, as those of the Infantry.
Quarter Masters	The Quarter Masters to wear crimson sashes round their waists.
Serjeants	The Serjeants to wear Pouches as the Men do, and a worsted sash about their waist, of the colour of the facing of the regiment, and of the stripes on the lace of the housings.

CAVALRY

1715–1760

For the cavalry of George I we have little pictorial evidence. There is one painting, however, which gives us some details of dress. This is the portrait of Captain William Pritchard Amhurst by Peter Tillemann. Amhurst, who joined the King's Own Regiment of Horse (K.D.G.'s) in 1722 and left the following year, is mounted on a long-tailed charger, with a stript saddle, consequently no housings are shown. He wears a plain red coat without lapels, silver buttons, large deep red cuffs, which reach to the elbows, and are

Fig. 79. Officer, King's Own Regiment of Horse, 1722.

ornamented with three silver stitched or laced buttonholes and buttons. The pockets are cut crossways and have four buttonholes; the breeches and waistcoat are yellow, and the latter is ornamented with three rows of buttons down the front and silver loops. These appear to be placed between three vertical strips of silver lace. The hat is bound with silver lace (Fig. 79).

In the background are small figures of troopers as well as trumpeters in yellow coats and mounted on grey horses. It is curious that the regiment which had changed its title from the Queen's Regiment of Horse to that of The King's Own in 1714, and is said to have changed its facings to blue at the same time, should be still wearing those of yellow.

An engraving of the funeral procession of the Duke of Marlborough shows a Troop of the Life Guards. The arrangement of their cuff lacing corresponds very closely with that of some troopers in one of the paintings at Marlborough House, the housings in the engraving having the crown and cypher within the garter, those in the painting being plain blue edged with gold lace.

The Trumpeters in the engraving wear three-cornered hats and coats with large cuffs and hanging sleeves.

The description of a review of the 4th Dragoons in 1715 gives us a complete picture of a regiment of the time. The writer says: "I scarce think there is a more showy regiment in Europe. They drew up in two squadrons, three lines in each squadron, which made six companies in all, each company had a farrier or hatchet man, with a very high black furred cap on, and a pock (i.e. bag or hood) hanging behind, tipped with fur. Instead of a sword they had a saw at their side and a shovel in place of their slinged carbine. They carried axes in their hands and a hatchet under their belt instead of pistols. There were fixed two large things likest to muff cases where their horseshoes and nails were; they have clean white aprons and white gloves, and rode upon good grey horses. The six drummers were Moors with brass drums, and the Hautboys and they rode upon grey horses. The six troops rode by the Duke (Argyll), troop by troop, with a hatchet man and the proper officers. Before every company there stood an officer where they were to wheel, when they came up five in a rank and to every file he said: 'Look the Duke full in the face.'

"The two squadrons when they passed had thus took up again their several grounds. The next they did was, dismounting and fixing all their horses, so that every tenth man held the nine horses, and then they drew out to that empty space betwixt the two squadrons and there they performed their exercise on foot. Then they marched on foot by the Duke, who had dismounted with them. They took up their ground again when they were past, retired in order to their horses and remounted and rode again in troops by His Grace, but as they passed him this time, every horse singly from the right to the left filed off and rode by, you might see the vanity of some to make handsome caracole, touch their horse to the quick with the spur. I never in my life saw so many fine black horses, the officers' led horses were extraordinarily fine, with fine decks of different skins to cover them. There was something very shog and neat in the appearance of the regiment, as footmen and the servants that led the deck horses."

The foregoing is interesting not only for the evidence that pride of regiment and smart turn-out were already strongly developed, but for the early notice of the use of markers and the original form of the order "eyes right." Besides this we have the earliest reference to fur caps, aprons and equipment of the farriers in mounted corps, and the enlisting of men of colour as drummers in the 4th Dragoons, a custom which this regiment continued up to the nineteenth century.

The decks of different skins for the led horses are to be seen in the mural paintings in Marlborough House, where a General's charger is shown with a covering of leopard skins.

Although this review took place in 1715 it is probable that the farriers' dress and equipment and the Moorish drummers had been adopted at a much earlier date.

In the beginning of 1715 we find the first mention of shoulder knots for mounted regiments. These, it has been suggested by some authorities, were originally the cords used when foraging to attach the hay, etc., to the saddle, hence the word *fouragére*, still used in the French army for the emblem which is granted to regiments which have distinguished themselves in action. It is suggested that troopers on setting out on a foraging expedition looped the cord over their shoulders as a convenient way of carrying it and that a custom started of colouring these cords according to regimental fancy until it finally became an ornamental part of their dress.* If this were so, it would

* This custom has been resuscitated in recent years by the wearing of lanyards of various colours, or combination of colours, of regimental pattern.

be very similar to the lanyard attached to the jack knife, which during the Great War was plaited by the men into various fancy patterns, although the knife itself was often carried on a waist-belt or in some other handy position.

It must, however, be pointed out that to support the above theory there are no examples shown in contemporary prints, or pictures, of troopers carrying anything resembling cords. Be this as it may, it had now become an ornament of dress attached to a button on the shoulder, and was plaited and hung in loops, the ends being furnished sometimes with metal points, very much like the present aiguilette.

The efforts to standardise the dress of the army were also applied to the cavalry, not that the orders seem to have been always strictly adhered to.

The following orders were issued:

"30 Oct. 1727. The King being desirous that all his regiments should have a fixed clothing and that each regiment should differ in their facings or otherwise, the several Colonels are to consult together to prepare patterns before they make up the next clothing, which they are to show His Majesty, and then orders will be given that the regiments continue that clothing and that they do not presume to make any alteration, whoever shall happen to come at the head of the corps.

"It is His Majesty's pleasure also that all Horse do wear cross belts of plain buff instead of ornaments of lace which several of them now have." (W.O. 71/6.)

The above reads as if rather more latitude was allowed in the dress of the cavalry. There is no mention that the coats are to be furnished with lapels. It was in any case the end, officially, of the laced belts with which regiments of Horse had enriched their equipment.

"1729. 31 Oct. For a Dragoon: New cloth coat well lined with serge, new waistcoat, pair of new breeches, new laced hat, pair of new large buff coloured gloves with stiff tops, pair of new boots, every four years to be given to the men at such times as the Colonel shall find it necessary within the four years, saddles left to the general officer inspecting. Housings and (holster) caps once in six years, or sooner if wanted, new horse furniture once in four years, bits and stirrup-irons as wanted, cloaks faced with the livery of the regt. entirely new once in six years, to be given to the men at such times as the Colonel shall find it necessary within the six years, new buff or buff coloured accoutrements, viz., a shoulder belt with a pouch, a waist-belt sufficient to carry the sword, with a place to receive the bayonet and sling for the arms once in ten years, or sooner if wanted.

"The 2nd. Mounting to consist of new laced hats, gloves and horse collars." (W.O. 71/6.)

"1736. 20 May. The orders are similar."

A set of these family portraits (Figs. 80 and 81) represent the uniform of a mounted corps of George I or early George II period. Although the hair dressing appears to be of different periods the uniforms are very similar. All are scarlet with silver lace and yellow cuffs; the latter are curious as they show a wide border of the scarlet of the coat cloth with the yellow of the cuff facing outside.

FIG. 80. Officer, early 18th cent.

The older officer, with his powdered wig, has plain scarlet lapels (Fig. 80); the younger men wear no lapels but a heavily silver laced yellow waistcoat. The sash of one has a central stripe of silver (Fig. 81).

FIG. 81. Officer of Cavalry.

In the fine painting by Dandridge of Captain Richard Gifford of the 4th, or Scots Troop of Life Guards (Fig. 82), he is not wearing the gold laced full dress but one of plain dull crimson velvet with cuffs of the same colour, buff laced waistcoat and breeches. These last-mentioned garments have fine embroidery around the button-holes. It may be pointed out that the coat is very similar in

FIG. 82. Officer, 4th Troop of Life Guards, *c.* 1735.

colour and style to that worn by Lord Charles Hay in his portrait. As there are other portraits of officers besides these wearing crimson velvet coats, one wonders if it were a fashion of that time to wear this type of garment when not in full regimentals.

The holster caps and housings are blue and deeply fringed and laced with silver; a cane hangs from the rider's wrist.

Captain Gifford is attended by his groom dressed in blue coat, buff waistcoat and wearing a huntsman's cap. In the background beside the Kettle drummer and Trumpeter, who will be described later on, are two troopers dressed in red coats and gold laced three-cornered hats.

The portrait is supposed to have been painted about 1735.

An engraving of the portrait of Captain James Miller, who served in the 4th Troop of Life Guards 1723–41, shows the full dress of an officer of the Troop (Fig. 83). The coat appears to be entirely of scarlet cloth, with the exception of the collar and lining. The lace looped and tasselled ends are interesting; they are also shown in the Morier painting of an officer of Horse Grenadier Guards, as well as in another painting of an officer of a hitherto unidentified regiment (p. 140), so that it is possible this style of lacing was typical of the Household Troops.

We now come back to our old friend the 1742 Clothing Book. Here the same system is followed as in the case of the infantry, that is, about six figures coloured differently are used to show the whole

FIG. 83. Officer, 4th Troop, Life Guards.

of the regiments of Horse and Dragoons. Starting with the Life Guards, one figure represents the four Troops, the only difference being the colour of the housings, holster caps and their carbine belts. These last have three bands of gold lace and two of the Troop colour, the 1st Troop's housings and carbine belts being red and gold, 2nd white (this in the book has changed through discoloration to a pinkish tone),* 3rd yellow, and the 4th blue.

The device on the housings and holster caps is the same for all, i.e. G.R. in gold on red

FIG. 84. 4th Troop of Life Guards, 1742.

* Millan's list gives buff.

within the garter, surmounted by the crown. The lace appears to be more in the form of fringes than the usual kind of lace (Fig. 84).

The uniform of all troops is a scarlet coat without lapels, but with gold lace loops either side of the front. Below the white neckcloth is a curious piece of blue cloth, edged with gold, which is buttoned to either side of the red gold-edged collar. The upper part of the cuffs is scarlet laced with gold, the remainder is hidden by the long buff gauntlet glove. The pocket flaps are laced with gold and the skirt turn-backs are blue. The waistcoat and breeches are buff, hats laced with gold, the cloaks red, lined blue. The swords are furnished with gilt basket hilts. The hair is powdered.

The 1st and 2nd Troops of Horse Grenadier Guards (and also the Royal North British Dragoons) are represented by a mounted figure carrying his firelock in the position of "advance your muskets." The Duke of Cumberland in 1745 ordered that they were always to appear so at review, and always to march past so.

The Horse Grenadiers wore cloth caps with the fronts embroidered with G.R. in yellow on a red ground within the garter and surmounted by the crown. The design on the little flap is indistinct.

FIG. 85. 1st Troop Horse Grenadiers, 1742.

Fig. 86. Regiment of Horse, 1742.

The 1st Troop, a red cap with blue front edged yellow, the little flap and turn up blue, and a blue tuft. The housings and holster caps are blue with the crown and cypher and garter and are edged with yellow lace with a red stripe. Blue flask cords (Fig. 85). [Millan's List gives red housings.]

The 2nd Troop, blue cap, red front embroidered in white and with the same devices as the 1st Troop, blue turn-up and little flap; this probably had the thistle, as it was the Scots Troop, red tuft. The flask cord is red and the housings, etc., are red edged with yellow lace with blue central stripe. The uniform, with the above differences, is the same, a red coat

Fig. 87. Trooper of Horse, 1742.

without lapels but with white lace loops and buttons down the front and a white shoulder knot. The waistcoats are buff with white buttonholes and the breeches buff.

The Life Guards fought at Dettingen and helped to cover the retreat after Fontenoy.

The eight regiments of Horse, and with these are included The Royal Horse Guards (The Blues), are represented by two figures, one a trooper firing his pistol and riding a prancing horse; the second, by one drawing his sword (Figs. 86 and 87).

CAVALRY

The Royal Horse Guards, The King's Own Regiment of Horse (1st K.D.G.'s), 5th Horse (4th D.G.'s) are represented by the first-mentioned figure (Fig. 86); the 3rd Queen's Horse (2nd D.G.'s, Queen's Bays), 4th Horse (3rd D.G.'s), 6th Horse (5th D.G.'s), 7th The King's Carabineers (6th D.G.'s The Carabineers), 8th Horse (7th D.G.'s) by the second figure (Fig. 87).

With the exception of the Royal Horse Guards, who have blue, all the regiments wear red coats with a row of buttons down the front. No lapels are shown. The hats are bound with white or yellow lace. This may be intended for silver or gold colour.

Take the first figure and note the different points: All three regiments have the same device on their housings, viz., the Royal Arms and supporters in full colours and a border of sprigs, or scrolls, which vary slightly in colour (Figs. 86 and 88 A).

In the second figure 87 again, the housings for all the regiments have the same device, a trophy of arms, standards, trumpets and kettle drums with a similar trophy on the holster caps; above this is a small wreath. The housings and holsters again have a border of scroll design (Fig. 88 C). In the case of the 4th Horse the housings are bound on the outer edge with a red lace, and leaves are introduced into the scroll embroidery (Fig. 88 B). The Carabineers also have leaves introduced as part of the design.

REGIMENTS OF HORSE, 1742

Regiment.	*Facings.*	*Waist-coat.*	*Breeches.*	*Cloak and Lining.*	*Flask Cord.*	*Hat Lace.*	*Housing & Holster Caps.*	*Embroidery on Housings.*
R. Horse Guards	Red	Red	Blue	Blue & Red	Red	Yellow	Red	White & Yellow
King's Horse	Blue	Blue	Blue	Red & Blue	Blue	Yellow	Red	Yellow & White
Queen's Horse	Buff	Buff	Buff	Red & Buff	Buff	Yellow	Red	Yellow & White
4th Horse	White	White	Red	Red & White	White	White	White	White & Red
5th Horse	Blue	Blue	Blue	Red & Blue	Buff?	Yellow	Yellow	Red & White
6th Horse	Green	Green	Green	Red & Green	Buff	White	White	White & Yellow
7th Horse	Yellow	Yellow	Yellow	Red & Yellow	Yellow & Buff	White	White	Red & Dark Blue
8th Horse	Black & Buff Linings	Buff	Buff	Red & Buff	Buff	Yellow	Yellow	Red, White & Blue or Black

A B C

FIG. 88. Housings of Regiments of Horse, 1742.

The 8th Horse have buff skirts and linings instead of black to match their facings. This regiment was known as Ligonier's from its Colonel John Ligonier. Major-General Ligonier was of a French Huguenot family, and under his Colonelcy the regiment attained a high reputation, and it was considered a privilege to be admitted to it; many Irishmen of good family served in its ranks.

During its repeated and brilliant charges against the Maison du Roi and the Gens d'Armes at the battle of Dettingen, Cornet Richardson became surrounded by the enemy, who attempted to capture his standard. This, however, he managed to preserve in spite of receiving upwards of thirty sabre cuts in his body and clothes, his standard and lance being also damaged in the fray. This standard was afterwards presented to him and was for long preserved in the family, until later it was presented to the Royal United Service Institute where it now hangs.

Ligonier was created a Knight Banneret on the battlefield by George II himself.

The fourteen regiments of Dragoons have two figures to show their dress, besides a separate one for the 8th. The Royal North British are represented by the same drawing as the Horse Grenadiers.

The 1st or Royal Dragoons, 3rd The King's, 5th The Royal Regiment of Ireland, 7th The Queen's, are all shown by the figure with his musket slung over his back (Fig. 89). The 4th, 6th, 9th , 10th, 11th, 12th, 13th and 14th by a Dragoon in the position of "Port arms" (Fig. 91).

In the first series the housings are all distinguished by the same device, the Royal cypher within the garter surmounted by the crown. The North British Dragoons also had this design (Fig. 90 A).

The remaining regiments all have a knight's helmet with trophies of colours, trumpets, drums and arms on their housings and holster caps with an elaborate border of embroidered scroll work. The 6th Inniskillings differed from the other regiments by having, besides a diamond pattern, an outer edging of red, yellow and black (Figs. 90 B and C).

The last regiments, 11th, 13th and 14th, present several details of colouring which may appear doubtful. They may be due to errors in colouring, or it may be that these regiments for some reason actually did have them.

FIG. 89. The Royal Dragoons, 1742.

The parts of the dress left white in the original are not due to any oversight, as they are actually painted in body colour. We have noted these divergencies as they are shown in the Clothing Book and the reader can compare them with Millan's Lists and with Morier's figures of the later date.

All regiments of Horse and Dragoons have flask-cords in the centre of their pouch-belts; some of these are of different colours. The Rev. P. Sumner says that only those with Royal titles have theirs coloured, but in the case of a regiment with yellow or buff facings it is not easy to say if the cord is coloured or left buff.

The Royal Dragoons seem to be the only regiment wearing their shoulder knot on the left side. Most regiments have a small patch of the facing colour on their red collars; this has a button and stitched loop. It may be that all regiments had this and that where it is not visible it is due to oversight on the part of the colourist. An actual coat exists in the Scottish National Naval and Military Museum, and an officers' cap in the Edouard Détaille

REGIMENTS OF DRAGOONS, 1742–45

Regiment.	*Facings.*	*Waist-coat.*	*Breeches.*	*Cloak and Lining.*	*Flask Cord.*	*Hat Lace.*	*Housing & Holster Caps.*	*Embroidery on Housings.*
1st Royal Dragoons	Blue	Blue	Red	Red & Blue	Blue	Yellow	Blue	White & Yellow
2nd Royal North British	Blue	Blue	Blue	Red & Blue	Red	Yellow	Red	Yellow & Green Foliage
3rd King's	Light Blue	Light Blue	Light Blue	Red & Light Blue	Blue	Yellow	Red	Yellow
4th ,,	Green	Green	Green	Red & Green	Buff	White	Green	White & Red
5th Royal Irish	Blue	Blue	Blue	Red & Blue	Buff	Yellow	Blue	Red & White
6th Inniskillings	Yellow	Yellow	Red	Red & Yellow	Buff	White	Yellow	Red & Black
7th Queen's	White	White	White	Red & White	White	White	White	Red & Blue & Yellow
8th ,,	Orange	Orange	Orange	Red & Orange	None	White	Orange	White
9th ,,	Buff	Buff	Buff	Red & Buff	Buff	White	Buff	Red & Black or Dark Blue
10th ,,	Yellow	Yellow	Yellow	Red & Yellow	Buff	White	Yellow	Red & Black or Dark Blue
11th ,,	White Cuffs, Buff Linings	White	Red	Red & Buff	Buff	White	Buff	Red & Black or Dark Blue
12th ,,	White	White	Red	Red & White	Buff	White	White	Red, Yellow & Black
13th ,,	Green Cuffs, Buff Linings	White	White	Red & White	Buff	White	Pale Buff	Red & Blue
14th ,,	Pale Yellow, White Cuffs	White	White	Red & Yellow	Buff	White	Pale Yellow	Red & Blue
Kingston's	Green	Buff	Buff	Red & Green	None	Yellow	Green	Yellow & Green Stripes

A B C

FIG. 90. Housings of Dragoon Regiments, 1742.

FIG. 91. Dragoon, 1742.

Collection, Musée de l'Armée, Paris, both of the Scots Greys (Figs. 92 and 93).

These two articles of dress agree with the drawing in the Clothing Book, which shows a red coat faced and lined blue, white shoulder knot, blue

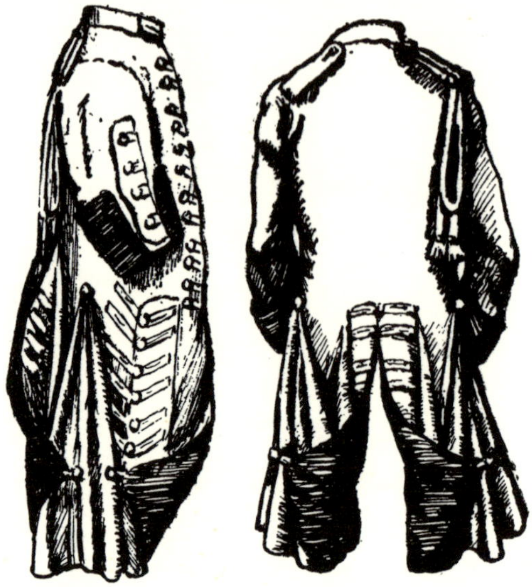

A B

FIG. 92. Coat, Royal North British Dragoon, *c.* 1743.

waistcoat and breeches with all buttonholes laced with white. The buff belts are whitened, this regiment being the only one to do so at this time. The flask cord is red. The cap is red, piped yellow, and has a red front embroidered in worsted with the same design as the officers', which is described below. The housings and holster caps are red and have the crown and garter devices with an embroidered pattern of foliage in green and yellow and the outer edges are bound with yellow lace. The cloak is red lined blue. The officer's cap has a red velvet front edged with gold. In the centre is the red Cross of St. George within the garter, which is blue worked with gold. This is all superimposed on two crosses, one described as Latin, worked in dark yellow, and the other St. Andrew's, worked in narrow strips of some thin white metal. Between these two crosses are rays worked in the same white metal. On either side of this are foliage, thistles and roses worked in silver or gold. The stalks and circular scrolls are worked in yellow. The little flap is blue edged gold and embroidered with a thistle, over which is a label worked in white with motto *Nemo me impune lacessit* in black. The back of the cap is plain red piped with gold and the turn-up is plain blue, but this I am told is reconstructed and not original. The embroidery on the front of the cap is of a warm silver colour, but it may have been originally gold (Fig. 93). Certainly some parts have a yellow tone. In 1737 we read of oilskin covers for the Grenadier caps. The regiment seems to have had both hats as well as caps.

A print of Thomas Brown of the 3rd Dragoons published by Carrington Bowles in 1743 shows him in a uniform very much like that shown in the Clothing Book. The device on the holster caps is the same, crown and garter, but within the latter is a shield with the Royal Arms instead of the cypher. A patch is shown on the collar and the coat has buttons and laced loops on both sides down the front; the waistcoat has loops on the left only. A powder horn, or flask, is visible. Another print of Brown published at the same time shows much the same details of dress.

FIG. 93. Officer's Cap, 2nd or Royal North British Dragoons.

Beneath these pictures is the story of Brown's exploit at the Battle of Dettingen, where two squadrons of the 3rd or King's Dragoons repeatedly charged ten times their number of French Household cavalry, losing two-thirds of their men and horses and having only two of their officers left.

Thomas Brown was a native of Kirkleatham in the North Riding of Yorkshire, "a tall raw-boned young fellow of twenty-eight years of age. He had two horses killed under him in the battle and was about to dismount to take up the standard (which had been dropped by the Cornet, who had been wounded in the wrist) when he had two of the fingers of his bridle hand chopped off by one of the enemy. His horse took fright and ran away with him to the rear of the enemy before he could stop him. In the meantime a Gen d'Arme had captured the standard. Brown pretended he was returning to his regiment, but suddenly made furiously towards the Gen d'Arme and shot him through the head with his pistol, the standard falling into his arms, upon which he clapped it between his leg and the saddle and galloped through a lane of the enemy. He then received eight cuts in the face, head and neck, two bullets in his back, and three through his hat, and in this hacked condition regained his regiment, who gave him three huzzas on his arrival." Brown was rewarded by being posted as a private gentleman in the Life Guards.

The 8th Dragoon (Fig. 94) in the different copies of the Clothing Book appears to have his facing, waistcoat and breeches and skirts either a brownish colour or a pinky red tone. The actual colour appears to have been an orange yellow as given in Millan's List. The housings are not visible, being hidden by the man "standing to his horse." This regiment had the privilege of wearing cross-belts like the Horse instead of carrying their swords in a waist-belt like the other Dragoons, and these are well shown on this figure and possibly supply the reason why he is drawn in a different pose to the other regiments. Apparently there had been proposals at various times to do away with the shoulder belt, which were vigorously resisted by the regiment. It was probably on account of some such suggestion

Fig. 94. 8th Dragoons, 1742.

being under consideration at the time that Morier has painted the 8th Dragoon with a waist-belt.

The origin of this distinction is explained by General Severne in one of his letters as follows: "When Colonel, now Lord Waldegrave, became Colonel in 1755 he had the Duke of Cumberland's order to take the cross-belts from the regiment. I then remonstrated against it and submitted my reason in writing to H.R.H. and we were permitted to continue to wear them."

General Severne, who was Colonel of the regiment from 1760 to 1787 and served in the Army from 1715, traced the origin to the Battle of Almenara, July 16th, 1710, where the 8th defeated and took prisoner a regiment of Spanish Horse, and when "the men exultingly took their prisoners' cross-belts and wore them as a mark of honour gained in battle, Marshal Starenberg permitted them to wear them."

General Severne again wrote in 1769: "By the King's order cross-belts were expressly confirmed to the 8th Dragoons, with an addition of carrying their swords like the regiments of horse, by their cross-belts, which before they carried in their waist-belts."

An order also states that "The 8th Dragoons are to continue their cross-belts, that over the right shoulder is for the sword and bayonet, they are therefore not to have waist-belts. It is so ordered by His Majesty and please inform the regiment as they are about preparing the waist-belts according to the former order sent to the Colonel." (W.O. 3/22.)

The following orders for a review held by Lord Stair on October 23rd, 1742, show the attention that was given to smart appearance and will tell their own story to any ex-cavalryman who may read them.

"The Horse and Dragoons to take all their small accoutrements to pieces, and to see that they be very well cleaned and blacked and then put together again. The bosses, bits, and curbs to be as bright as hands can make them. The boots to be as black as possible, and their knee pieces not to appear above 3 inches above the boot top. All their arms are to be as bright as silver. The whole buff accoutrements to be of one light buff colour, the swords to be all brightened. The hats new cocked, 3 straps to each cloak, care to be taken that the men do not ride too long. Officers to wear their sashes over their shoulders."

The knee pieces mentioned above, and in other places called canvas boot stockings, were pieces of white cloth or canvas, 8 inches long, which

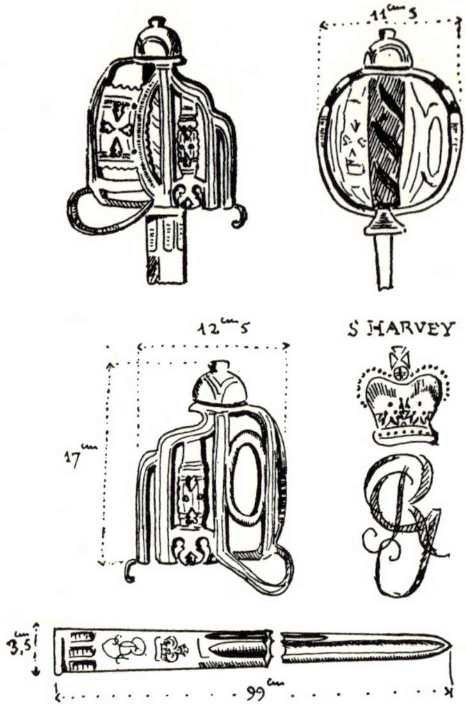

FIG. 95. Cavalry Sword, *c.* 1758.

buttoned over the knees to preserve the breeches and to protect them from the blacking on the boots. They were tied at the base with a string and at the top were fastened to the button on the knee of the breeches. In some of the Clothing Book figures the top is fastened by a white, or sometimes a yellow, tag to the breeches.

Other regimental orders mention that "the men are always to appear clean and tidy in the streets with their hair under their hats, good stockings and shoes and uniforms clean."

Dragoons also had white or black gaiters for dismounted duties. The officers' uniforms were laced with gold or silver and their sashes worn over the left shoulder. The housings would be laced gold or silver and would have the knot and tassel on both housings and holster caps.

The Quartermaster wore a crimson sash round the waist, the Sergeants one of worsted with a stripe of the facing colour, one order stating that all regiments of Horse Grenadiers and Dragoons were to wear their sashes.

The size of the horses was:

For a Trooper of Horse a strong well-bodied animal from 15 hands 1 inch to 15.2.

For a Dragoon a strong well-bodied horse of 15 hands.

It is obvious that, except for a slight disparity in the size of their horses and the arrangement of their accoutrements, there was now practically no difference between Horse and Dragoons. They were, in fact, what at a later date were called Heavy and Medium Cavalry.

The Dragoons seem now to have become quite unsuited to their original duties, which were those of light horse. The lack of light cavalry was greatly felt in the campaigns on the Continent, where foreign corps were only the ones available to cope with the Hussars and Troupes Legéres of the enemy.

During the Jacobite rising of 1745 two regiments of cavalry were formed, Montague's Carabineers and Kingston's Light Horse.

Sewell, in his article on Disbanded Regiments in the *R.U.S.I. Journal*, says that the Dukes of Montagu, Bolton and Ancaster, the Marquis of Granby, the Earl of Berkeley, Viscount Falmouth and Viscount Harcourt raised regiments of Horse besides *their* regiments of Foot. He also mentions a Corps called the Derby Blues.

FIG. 96. Cumberland's Dragoons, 1746.

"The Duke of Montagu raising a regiment of Horse in Northants of gentlemen, tradesmen and tenants, clothing to be Blue lin'd red, and yellow buttons." (The *Ipswich Journal*, 7th March, 1744.) One wonders what this regiment was; the date seems to be too early for it to have been raised owing to the Jacobite scare.

Kingston's Light Horse was raised by the Duke of Kingston at his own expense upon an entirely new plan to imitate the Hussars in Foreign service. Hinde, in his *Discipline of Light Horse* has a good deal to say about them. He tells us that "they were mounted on light horses of various colours with swish or nick'd tails. Their accoutrements were as light as possible of every sort and species. Their arms were short bullet guns, or carbines, shorter than those of regiments of Horse and slung to their sides by a movable swivel to run up their shoulder belt: their pistols upon the same plan, as they used both carbines and pistols on horse-back, indiscriminately. Their swords were very sharp and inclined to curve."

The regiment did good service in the campaign and distinguished itself at Culloden and was disbanded at Nottingham, September 15th, 1746, and another regiment of Dragoons raised in its place—"at the same time and place where your Grace's regiment shall be disbanded . . . and as a signal mark of honour and distinction H.R.H. the Duke of Cumberland will himself be Colonel." (W.O., Sept., 1746.)

The men of Kingston's on being asked to re-enlist, "all men excepting eight entered immediately into the D^{k} of Cumberland's new regiment, and those gentlemen who did not enter gave reasons very satisfactory and honourable." It is further mentioned that there were three butchers of Nottingham, that had entered into the Duke of Kingston's regiment, who killed fourteen rebels each at the late battle of Culloden.

The regiment was now raised again from Horse to be Light Dragoons, but mounted on the same sort of light nag-tailed horses from 14½ to 15 hands high. Their arms and accoutrements, furniture, saddles, bridles, and all their appointments entirely on a heretofore new and light plan. The size of the men from 5 ft. 8 ins. to 5 ft. 9 ins., but wore hats as the heavy Dragoons, and not helmets. Hinde mentions this point about the headdress, forgetting that helmets had not yet become a distinctive article of dress for Light Dragoons.

The regiment went to Flanders under the command of H.R.H. the Duke of Cumberland and served with distinction at the battle of Val or Kistelt, July 2nd, 1747. This was the Battle of Lauffeld where the regiment, together with the Scots Greys and Inniskillings, took part in the desperate charge led by the gallant Ligonier in person to cover the retreat of the infantry, Ligonier himself being overthrown and taken prisoner.

At the Peace of 1748 the regiment was entirely disbanded and no light horse were retained until just before the war of 1756.

The uniform is shown in one of Morier's paintings, being a red coat with brass buttons and yellow buttonholes, green cuffs and a green oval patch edged yellow on the red collar, yellow shoulder knot and green skirt turn-backs, yellowish buff breeches, waistcoats and gloves. The hat is bound with yellow or gold coloured lace and has a green plume, the first instance

FIG. 97. 1st Horse (4th D.Gs.), 1751.

of a feather worn in the three-cornered cocked hat, green and red cloak, green housings and holster caps edged yellow lace with green stripe in the centre, the crown over the garter in the centre of which is the crowned lion of England.

The 8 Drummers are described in the *Ipswich Journal*, 1746, as wearing scarlet coats with green facings and gold lace, the Duke's livery.

The uniform, when Kingston's regiment, seems to have been much the same, as it was changed from one regiment into the other without a break, the exception being made in the case of the crowned lion which was a royal badge. Millan's List, 1749, gives green facings, yellow lace and green horse furniture under the title of Kingston's Horse.

In December, 1746, for reasons of economy, the first three senior regiments of Horse after the Royal Horse Guards were transformed into Dragoons (the pay of Dragoons at that time being less than that of troopers of Horse) and as a compensation the title of "Guards" was prefixed. In consequence of this the 4th, 5th, 6th and 7th Horse now became respectively the 1st, 2nd, 3rd and 4th. For some reason, at the same time, the title of the King's Carabineers was changed to the 3rd Irish Horse.

By consulting the Clothing Warrant and Morier's paintings it will be seen that the dress of the cavalry was now based on a regular system.

FIG. 98. 3rd Irish Horse (6th D.Gs., The Carabineers), 1751.

It will be easier for our purpose to quote the essential articles of the Warrant.

All regiments with the exceptions noted below were to have cuffs, linings and lapels, waistcoats and breeches of the facing colour as well as their housings and holster caps.

The regiments of Horse were to have coats lapelled to the bottom. Their cuffs were round and without any slash or slit. No shoulder knots are shown. The 4th Horse have black cuffs and lapels, but skirt linings, breeches, waistcoat and housings of buff.

Dragoon Guards. Their coats are to be lapelled to the waist with slit cuffs.

FIG. 99. 2nd Dragoon Guards (The Bays), 1751.

Dragoon Guards and Dragoons to have shoulder knots of white or yellow worsted worn on right shoulder. The whole to have long pockets, the buttonholes to be of narrow yellow or white lace and set on in twos or in threes.

The coats of Dragoons are in some cases described as double-breasted, and the Order, Dec. 4th, 1752, describes what is really meant, viz.: "All coats of Dragoons to be double-breasted, whether they understand by this, buttons and holes on both sides, or only buttons on one side, and holes on both, is equal to H.R.H. (Duke of Cumberland) provided the clothing of all Dragoon Regiments be alike in it." (W.O. 7/25.)

In 1753 (Jan. 8th), it was found that all regiments of Dragoons except the Royal North British and King's Own Regiment (3rd Dragoons) were

FIG. 100. Officer's Cap, Royal North British Dragoons, *c.* 1751.

clothed alike. The two regiments named wore buttons and holes on each side and they were ordered to follow suit to be uniform.

Sergeants of Dragoon Guards and Dragoons had their coats laced with gold or silver with shoulder knots of the same, Corporals of Horse were similarly distinguished. Corporals of Dragoon Guards and Dragoons also had gold or silver lace, but with yellow or white silk shoulder knots. All the coats have red collars and a patch of facing colour with a button and loop. The cloaks were red with lining and cape (collar) of facing colour, and had loops of the same colour as the lace of the housings.

The watering, or forage cap, was red turned up with facing cloth and the rank of the regiment on the front. A small figure in the background of one of Morier's groups of infantry is depicted wearing a cloth cap with hanging hood or bag. It is very similar to that worn by French Dragoons at that date, which developed into the *bonnet de Police*, as the forage cap of a later period was called. (Fig. 100).

Kettle Drummers, Trumpeters (of Horse), Drummers and Hautbois (of Dragoons) wore coats of facing colour, lined and turned up with red and laced with the same coloured lace as on the housings; red breeches and waistcoats.

FIG. 101. 13th Dragoons, 1751.

The Royal regiments, however, wore Royal Livery, i.e. red lined blue with blue waistcoats and breeches. All had long hanging sleeves.

The cloaks were red cloth with capes or collars of the facing colour and lined like the coats. They were fastened down the front with buttons and frogs or loops of the same colour as the lace of the housings. Forage or watering caps were red turned up with the facing colour and the rank of the regiment on the little flap.

Sergeants of Dragoon Guards and Dragoons were distinguished by a narrow gold or silver lace on the lapels (for Dragoon Guards), turn-up of sleeves and pockets and had gold or silver shoulder knots; Corporals of Dragoon Guards or Dragoons by a narrow gold or silver lace on the cuffs and shoulder straps and had white or yellow silk shoulder knots.

Quartermasters wore crimson sashes round the waist and Sergeants worsted waist sashes of the colour of the facing and of stripes of the housing lace and were to wear a pouch like the men. It may be remarked that the few pictures of Sergeants show a crimson sash with a stripe of the facing colour.

FIG. 102. Farrier, 11th Dragoons, 1751.

FIG. 103. Officer 8th Horse, *c.* 1745 (7th D.Gs.)

The Farriers wore a blue coat (Fig. 102). The only pictures we have are two paintings by Morier, both of the 11th Dragoons. These show single-breasted blue coats with white buttonholes, buff cuffs and skirt lining with a red collar patch, buff waistcoat bound and looped with white and buff breeches; a cap of dark brown fur with red bag with a white tassel. The front is ornamented with a red plate either of painted metal or embroidered. It is edged white and has a horse shoe with a hammer and pincers on either side in white. The housings are buff and like those of the rank and file; they appear, however, to have been decorated sometimes with the hammer and pincers. Farriers carried axes in a leather case and slung by a strap over the left shoulder. When the regiment was ordered to draw swords the Farriers "ported axes" in just the same way as is done in the Household cavalry to-day. On the front of the saddle a churn was attached and in this were the various tools, shoes, nails, etc. In some regiments the Farriers were mounted on horses of a different colour to those of the Troop. The leather apron is seen tucked up round the waist when the man is mounted.

Officers' uniform was to be like that of the men, but laced or embroidered with gold or silver; the waistcoats and breeches to be of the same colour as those of the men; but this does not seem to have been always followed, as they frequently appear to have worn breeches of scarlet cloth, even in those cases where those of the rank and file were of another colour. The crimson silk sash was worn over the left shoulder and a gold or silver aiguilette on the right. Their housings were of the same colour as the rest of the regiment, but edged with a gold or silver lace with a central stripe of velvet the same colour as the men's. Their housings and

FIG. 104. Officer, Horse Grenadiers, *c.* 1750.

holster caps, instead of the regimental device or number, were usually ornamented with a knot and tassels as distinguishing commissioned rank; sword knots of crimson and gold in stripes like the infantry.

The portrait of Ligonier as Colonel of the 8th Horse (7th D.Gs.), *c.* 1745, gives a good example of a cavalry officer's dress and horse furniture. The black velvet lapels laced with gold reaching to the bottom of the coat are clearly shown. The waistcoat is buff and laced with gold, but the breeches are scarlet instead of buff as worn by the regiment. The housings are red instead of buff, heavily laced and fringed with gold and the holster flounces are ornamented with the knot and tassels. The horse trappings are red and gold with a profusion of gilt crescents. The bridle and reins are red with gold and brass ornaments and the brow band is furnished with red silk rosettes (Fig. 103).

Another portrait also attributed to Morier is that of an officer of the Horse Grenadiers. The horse with its trappings is almost identical with those in the Ligonier portrait, the housings and holster caps are red and gold, and the tassel and knot are again carefully indicated. The bridle is red and ornamented with the same pattern of brass or gilt crescents, etc. The waistcoat is buff laced with gold and the breeches buff. The coat is single-breasted and has gold embroidered loops on the front and on the blue cuffs (Fig. 104).

A portrait of an officer of the 2nd Troop of Life Guards shows him wearing a scarlet coat with blue lapels; the buttonholes are embroidered with gold; plain blue cuffs and red slash with embroidered buttonholes (Fig. 105).

This is so very different from the usual nearly gold-laced coat that it is probably the frock clothing, used for ordinary duties. This officer only served with the regiment for a short time and left in 1757.

FIG. 105. Officer 2nd Troop of Life Guards, 1756–58.

The portrait of the Marquess of Granby by Sir Joshua Reynolds (Fig. 106) shows him as Colonel of the Royal Horse Guards in a blue coat with red cuffs and lapels laced with gold. The waistcoat and breeches are buff and the former is bound with gold lace at the edges and pocket flaps. The Marquess wears a cuirass under his coat and a sash round the waist and not over the shoulder as one would expect. This, it has been suggested, was because he was a General Officer, the sash being worn in that position by officers of that rank.

FIG. 106. Officer, Royal Horse Guards.

Morier's painting of a trooper of the regiment wears a blue coat with lapels, cuffs, waistcoat and breeches of red; there is no lace on the lapels. A powder flask is shown carried on a red cord. The housings are red edged with gold lace.

A very rare specimen of an officer's coat is to be seen in the Royal United Services Museum (Fig. 107). It is attributed to the 3rd Horse (The Carabineers 6th D.Gs.), and is of a crimson cloth with cuffs and half lapels of yellow richly embroidered with a graceful foliage design in silver. One wonders if the leaves are intended to represent shamrocks as they are certainly trefoil

FIG. 107. 3rd Irish Horse (6th D.Gs.), detail of embroidery on Cuff and Pocket of Officer's Coat.

in shape. It was at this time entitled the 3rd Irish Horse; that of "Carabineers" seems to have been dropped for a short while, at any rate officially. It is curious that the half lapels of a Dragoon Guard are given to the coat, instead of lapels all the way down as one would expect for a regiment of Horse. Still, in each one of these examples of officers' dress there are details which do not agree with the regulations of the Warrant and it seems evident that officers' dress was, as yet, by no means standardized.

FIG. 108. Officer's Coat, 3rd Irish Horse (6th D.Gs.)

The Life Guards being Household troops do not appear in the Warrant, but Morier depicts two of the 1st Troop and one of the 2nd.

Those of the 1st Troop (Fig. 108), although wearing much the same uniform, show several minor differences, the chief being that whereas one has the gold lace loops down the front of his coat at equal distances, the other has his in pairs. The pocket flaps of both are laced with gold, but in somewhat different ways. The uniform of both is a scarlet coat without lapels, scarlet collars edged with gold lace, with a blue patch on front edged with lace, and laced buttonhole loop, blue cuffs with scarlet slash, both heavily laced. The skirts are blue edged with lace, one having a gold loop, the other being without, crimson carbine belts laced with gold, the waist belt of one being of blue and gold, the other is hidden by the position of the arm. Buff waistcoats, one being edged with lace and having laced buttonholes; buff breeches for both. The housings are of scarlet edged with two gold laces showing scarlet between. The crown over the garter with the cypher in gold on scarlet of the housings. The figure of the second troop is much the same with the following differences: the laces on the front of the coat are in pairs, plain blue skirts, buff waistcoat and breeches, the former laced with gold down the front and on pocket flaps. The housings as the others. The sword hilts of both troops are brass or gilt, but of different patterns. The carbine belt is blue instead of crimson and is laced with gold.

FIG. 109. 2nd Troop Life Guards, 1751.

FIG. 110. 2nd or Scots Troop Horse Grenadiers.

FIG. 109A. 2nd Troop H.G.

FIG. 111. 1st Troop Horse Grenadiers, *c.* 1750.

Morier has also painted a negro Trumpeter of the Household cavalry, but beyond the fact that he wears a three-cornered hat, his dress is still worn at the present time with the exception that he has possibly hanging sleeves, but these from the position of the figure are not distinguishable. The housings are scarlet and gold like the Troop. (See Colour Plate).

The 1st Troop of Horse Grenadiers. The cap is blue fronted, embroidered with the crown and cypher within the garter and white scrolls or foliage. The little flap is red edged yellow and has what appears to be a grenade, but this is not clear in the original painting. The cap itself is red with a white tassel;

FIG. 112. 1st Troop Horse Grenadiers, *c.* 1750.

the turn-up at the back is also red edged white. A dismounted figure shows the back of the cap. The main body of the cap is not stiffened but hangs down with a large white tassel; the red turn-up has what looks like a drum with a crossed sword and musket either side embroidered in white. The powder horn with its blue flask cord is well shown as well as the white shoulder knot (Figs. 111, 112, 113 and 114).

The red coat is without lapels and laced with white. The cuffs of the dismounted figure seem to be of a different pattern from others. The waistcoat and breeches are buff.

The housings are blue edged with a double gold lace with a red central stripe; the crown, cypher and garter as before.

The 2nd Troop, or Scots Troop, differs by having the front of the cap red edged white and has the crown and garter as the 1st Troop, but the embroidery in this case very clearly shows the rose, thistle and foliage on either side. The little flap is blue edged white with a large white thistle having a yellow scroll with *Nemo me impune lacessit*. The turn-up at the back is blue edged white and shows the crossed sword and musket. The flask cord is red. The rest is similar to the 1st Troop, the shoulder straps and collar being red in both cases and the cloaks red and blue. The hair is depicted as being powdered on all the figures. The housings of the 2nd Troop are red edged yellow with a central stripe of blue, with a narrow outer edge of red appearing beyond the lace. The sword hilt is steel and the scabbard black leather. These flask cords are still worn on the pouch belts of the Household cavalry to-day (Fig. 110).

A long series of oil paintings in Windsor Castle show gentlemen of the 2nd Troop of Life Guards performing the musket manual exercises. The date of these is not easy to determine, as certain details of dress do not agree with other known pictures. Their whole appearance seems to be of the

FIG. 113. 1st Troop Horse Grenadiers.

fuller uniform and the heavier boots as shown in the 1742 Clothing Book, to which they bear a closer resemblance than to the Morier paintings. They may be possibly of the time of George I. The deep blue cuffs are edged top and bottom with gold. This is probably an error, but has been left as in the original. There is a narrow slit showing the red of the sleeve and the gold edging finishes in three loops. The collar is scarlet edged gold with gold loop and button but without any blue patch. The shoulder strap is gold laced on blue. The front of the coat has gold laced buttonholes in pairs with gilt buttons on the outer ends. The skirts are blue with gold loop and button, and the back of the skirts show gold loops. The waistcoat is buff bound with gold lace. The waist-belt and sword frog are also buff with

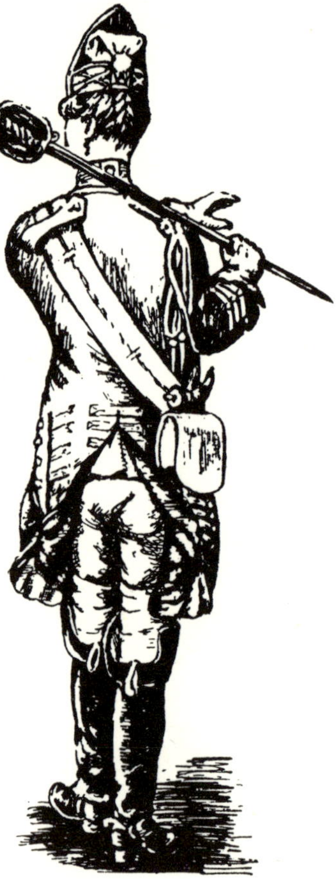

FIG. 114. 2nd Troop Horse Grenadier.

three gold laces. The breeches are buff. The carbine belt is blue laced with gold. The pigtail is well shown. This was not worn by ordinary private soldiers, but the Household cavalry being composed of gentlemen troopers, they would be quite likely to wear their hair in the same fashion as the officers (Figs. 114a, b. c).

The sword guard is of steel and lined with buff. These figures are interesting as they show side and back views of the uniforms and give details which would otherwise be mere speculation.

Ligonier for years past had advocated the re-issue of cuirasses to all regiments of Horse, Dragoon Guards, and Dragoons, between whom there was now little difference except in name and pay, as they were all practically heavy horse and acted as such.

In 1758 after they had been laid aside since Marlborough's time they were to be re-issued to the Royal Horse Guards (the Blues).

"5th July, 1758. The Royal Regiment of Horse Guards being ordered on

FIG. 114A, B. Life Guards.

FIG. 114C. Life Guard.

immediate Foreign Service and it being necessary they should immediately be provided with a complete set of cuirasses and scull caps, which they have not time to provide in the ordinary way, I desire you will acquaint the Board of Ordnance that it will greatly facilitate the service if they will give directions for them being furnished with complete sets of cuirasses and scull caps from H M stores, on their being replaced by others or paid for by the Royal Regiment of Horse Guards as the Board of Ordnance shall think fit.

"(Signed) Barrington."

In 1760 the 3rd Horse (6th Dragoon Guards) and 4th Horse (7th Dragoon Guards) were also provided with cuirasses and sculls. Cavalry were still formed in three ranks and wheeling by troops instead of each man turning his horse on its own ground was now the practice.

All the horses of the heavy cavalry are shown as blacks in the 1742 Clothing Book, except the Royal North British Dragoons. Morier's paintings show the same type and coloured horse, except that the 2nd Dragoon Guards are mounted on the bays from which they were to be known as the Queen's Bays. The tails are shown in the Clothing Book and Morier's pictures as cut extremely short.

The approach of the Seven Years' War found the army without any light cavalry. Cumberland's Dragoons had been disbanded in spite of its services and the vigorous efforts of the Duke for its retention. To remedy this deficiency orders were issued in January 1756 for the raising of light troops, the Warrant of April 14th 1756 reading: "A Troop of Light Dragoons to be added to each regiment of Dragoon Guards, and Dragoons, on the British Establishment of: 1 Captain, 1 Lieut., 1 Cornet, 1 Quartermaster, 2 Sergts., 3 Corporals, 2 Drummers, 60 Privates, including a Farrier. Height 5 foot 6½ inches to 5 ft. 8 in. To be light active young men.

"Horses 14 hands 3 inches, and not under, to be well turned, nimble, road horses, as nigh to the colour of the horses of the regiment as can be got.

FIG. 115. 8th Dragoons, 1751.

"Arms: Carbine, with ring, and bar 4 ft. 3 inches long, with a bayonet of 17 inches, 1 Pistol 10 inches in the barrel, and of carbine bore. Straight cutting sword, 34 inches in the blade, with a light hilt without a basket.

"Accoutrements: Tanned leather shoulder belt 3¼ inches broad, with spring and swivel, tanned leather belt for sword and bayonet, tanned leather cartouch box, with a double row of holes to contain 24 cartridges, with a tanned leather strap $1\frac{3}{4}$ inch broad. The saddle to be with small cantles behind as the jockey saddles are and to be 22 inches long in the seat; on the right side of the saddle is to be the holster for the pistol, and on the left a churn, in which a spade and felling aze, or a spade and woodmans bill is to be carried. There is to be a bucket for receiving the butt of the carbine, and a pipe to receive the end of the horse picket.

"The bridle to be plain, but with single reins; there is likewise to be a tanned leather head stall with a hempen collar.

"The clothing and cloaks of these light Dragoons to be as that of the rest of the regiment, only instead of hats the men are to have jockey caps ornamented in the front with H.M.'s cypher and crown in brass, and the number or rank of regiment. The crest is likewise to be covered with brass, out of which is to be a tuft of stiff horsehair, coloured half red and the other half of the facing colour."

Fig. 116. Officer's Cap, Light Troop 2nd or North British Dragoons, *c.* 1756.

The term jockey cap is somewhat misleading for what was in reality a helmet of black japanned boiled leather with metal mountings and in some cases even entirely of brass, or some other metal, as we shall see later on, that worn by the Scots Greys being different in having the embroidered cloth front and little flap similar to the Grenadier cap of the rest of the regiment. The plume on the specimen in the Royal United Services Institution is a drooping one of red, white and blue horse hair, unless, of course, it was originally tied or bound up so as to be of a stiffer shape (Fig. 116). Light jockey boots with small stiff tops, horse furniture of the same colour and form as the rest of the regiment and ornamented in the same manner but to be less in proportion to the size of the horses. The regiments which had light troops added to them were: the 1st King's Dragoon Guards, 2nd Dragoon Guards, 3rd Dragoon Guards, 1st Royal, 2nd North British, 3rd King's Own, 4th, 6th Inniskilling, 7th Queen's, 10th and 11th Dragoons.

These Light troops seem already to have been considered as something of

FIG. 117. Light Troop 11th Dragoons: laces were in threes.

a picked body of men and to have undergone a special training, which is described in the *Weekly Journal*, May 23rd, 1758, as follows: "The hussars (i.e. Light Troops) of the 9 regiments are now preparing to go on the expedition (i.e. to the coast of France). The flower of these Hussars is the Troop commanded by Capt. Lindsay (Light Troop 11th Dragoons) quartered at Maidenhead where they have been practising the Prussian exercise and for some days have been digging large trenches and leaping over them, also leaping high hedges with broad ditches on the other side. Their Captain on

Saturday last swam with his horse over the Thames and back again and the whole Troop were yesterday to swim the river."

These Prussian exercises were no doubt based on the cavalry training of Frederic II.

The pitiful display given by the Prussian cavalry at Mollwitz in 1741 brought from Frederic II an outburst of contempt and abuse; he called them "colossi on elephants, who can neither manœuvre or fight—are slow and before the enemy are no use and are always too late." This came strangely from the King who had himself fled ignominiously from the field while the fate of the battle was still undecided.

After this action every effort was made to reorganise the training of Frederic's cavalry, and the regulations of 1742 and 1744 give endless orders on charging at the gallop and stressing the fact that every trooper should be trained to ride "en écuyer" and jump ditches singly and in sections.

It was this Sovereign, who was acclaimed by our incurable Germanophiles of those days as "The Protestant Hero", quite oblivious of his saying that he "did not care what rubbish the fools sang from their hymn books."

The caps of drummers were like those of the infantry with the tassel hanging behind. The front was of the facing colour with the badge of the regiment or a trophy of guidons and drums. The little flap was red with the white horse and *Nec aspera terrent* round the edge. The back was red with the turn-up of the facing colour and embroidered in the centre with a drum and rank of the regiment. (*Vide* Frontispiece).

The Kettle Drummers and Trumpeters wore hats bound with the same lace as that of the regiment.

The Drummers of the 4th Dragoons were blacks and no doubt wore turbans.

Drummers of Dragoons usually carried scimitar swords, which appear to have been merely curved sabres.

The banners of Kettle Drums and Trumpeters were of the facing colours with the badge or rank of the regiment in the centre of the Kettle Drum banner as in the 2nd standard. The King's cypher and crown to be on the Trumpet banners with rank of the regiment underneath.

Drums of Dragoon Guards and Dragoons to be of brass, the front painted the regimental facing colour and upon which to be the rank or badge of the regiment as in the 2nd colour.

In the picture of the 1st Horse the Kettle Drummer has a pale or sky-blue coat laced with red and white lace, the knot on the front below the cypher is clearly shown. This, it will be remembered, was described in the Wardrobe accounts. (*Vide* Frontispiece).

CAVALRY KETTLE DRUMMERS AND TRUMPETERS

1714–1760

The Kettle Drummers and Trumpeters of Horse and Drummers, and Hautbois of Dragoon Guards and Dragoons followed the same regulations as the Infantry for their uniforms in having coats of the facing colour lined and faced with red, with red breeches and waistcoats. The coats were to be laced with the same coloured lace as that on the housings.

Royal regiments wore the Royal livery, red lined and turned up blue with blue waistcoats and breeches.

The Kettle Drummers and Trumpeters of Horse and the Drums and Hautbois of Dragoon Guards were to have long hanging sleeves. The Dragoons, judging by the paintings, also had these hanging sleeves.

The Kettle Drummer and negro Trumpeter in the background of the portrait of Captain Gifford, 4th Troop of Life Guards 1735, already mentioned on page 112 (Fig. 82), are invaluable as being the earliest evidence of their wearing Oriental dress. The owner of the painting, Mrs. C. M. Vivian Neal, has most kindly given the following description of the colours of their dress:

"The Kettle Drummer seems to be brandishing a staff similar to a Drum-Major's in his left hand and drum-stick in his right. He wears an oriental head-dress with a dark blue crown, which is turned up with yellow (i.e. the turban) and has a wide cockade of a light colour. His coat, waistcoat and breeches are red, as are also those of the Trumpeters, but the hats of these are the ordinary black tricorn hats bound with gold lace. The Trumpet banner is nondescript in colour and design."

The negro Trumpeters' coats seem to be similar to that shown in the Morier painting in Windsor Castle.

The Trumpeter (Fig. 118) is from a painting in the possession of Major Clarence Goff, who has kindly allowed me to make this illustration of it. The coat is scarlet bound with gold lace round the neck and down the front. It is further ornamented on the front, sleeves and pockets with

FIG. 118. Trumpeter in Royal Livery, *c.* 1735.

bands of gold lace edged with bright blue. The cuffs are of a darker blue with gold and lighter blue lacing. There seems to be an indication that the coat may also have been furnished with hanging sleeves. The waistcoat is dark blue bound with gold lace on the edges, pockets and button-hole loops.

The belt and gloves are buff. The sword has a white metal guard. The stock is white. The Trumpet is silver with crimson, blue and white cords and tassels. The hat is edged with gold lace. The man's hair is fair and not curled at the sides (Fig. 118).

The Lady Cecilie Goff has kindly given the following interesting notes concerning this painting: "The picture came from Windsor Castle. When my husband's grandfather, Lord Augustus FitzClarence, was given the house of Maple Durham on the accession of William IV Lord Augustus was given furniture, etc., from Windsor. In Lord Augustus's catalogue of 1845 the picture is described as 'A Trumpeter of George II by Dahl'."

Although this painting undoubtedly represents the Royal Livery, the arrangement of the lacing is quite different from that shown in other pictures of Trumpeters of the Household Cavalry (*vide* Frontispiece). It seems probable therefore that this Trumpeter belongs not to the Household Troops but to a regiment entitled to the privilege of wearing the Royal Livery.

CAVALRY: APPENDIX I

FACINGS AND HOUSINGS FROM MILLAN'S LIST. Published 1749

HORSE

1ST TROOP OF LIFE GUARDS, blue, red furniture, shoulder knots.
2nd Troop blue, buff furniture, shoulder knots.
3rd Troop blue, yellow furniture, shoulder knots.
4th Troop (Scots) blue, blue furniture, shoulder knots.
1st Troop Horse Grenadiers, blue, red furniture.
2nd Troop Horse Grenadiers (Scots) blue, blue furniture.
ROYAL HORSE GUARDS, red, red furniture.
2ND KING'S OWN ROYAL Regiment of HORSE, blue, red furniture.
3RD QUEEN'S OWN ROYAL Regiment of HORSE, buff, red furniture.
4th, White, furniture white.
5th, blue, blue furniture.
6th, green, green furniture.
7th King's Carabineers, buff, buff furniture.
8th, black, buff linings and furniture.

DRAGOONS

1st Royal Regiment, blue, blue furniture.
2nd Royal North British, blue, blue furniture.
3rd King's Own, pale blue, furniture red.
4th, green, furniture green.
5th Royal Irish, blue, furniture blue.
Inniskilling, yellow, furniture yellow.
7th Queen's, white, furniture white.
8th, orange, orange furniture. 9th, buff, buff furniture.
10th, yellow, furniture yellow. 11th, buff, furniture buff.
12th, white, furniture white.
13th, yellowish green, furniture buff.
14th, yellow, furniture none given.
Kingston's Light Dragoons, green, yellow buttons, furniture green.

The Life Guards, Royal Horse Guards, King's, Queen's, 5th and 8th regiments of Horse and Royal Dragoons, 3rd King's, Royal Irish have brass buttons and all other regiments white metal.

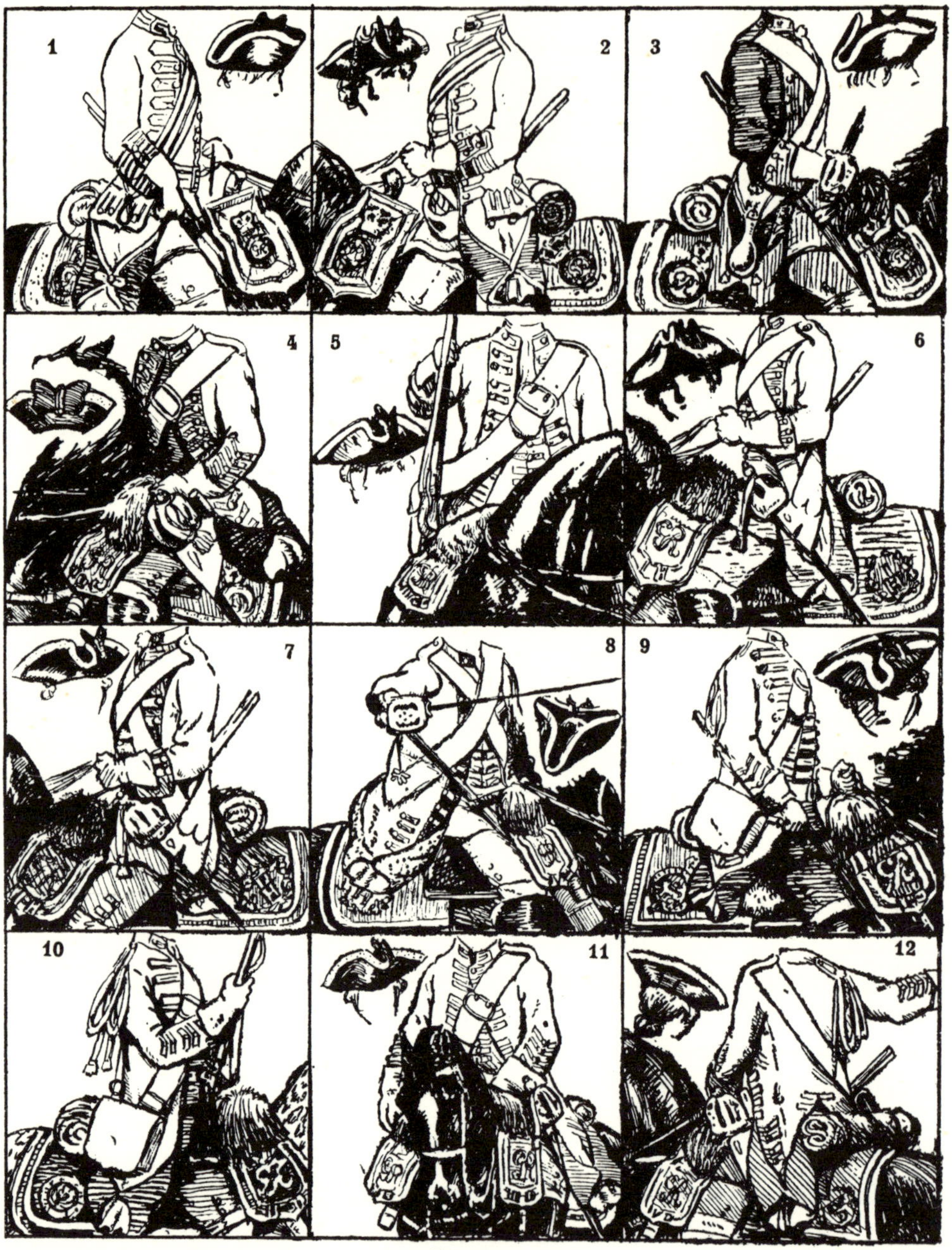

FIG. 119.

(1 and 2) Life Guards. (3) Horse Guards. (4) 1st or King's Dragoon Guards. (5) 3rd Dragoon Guards. (6) 1st Horse. (7) 2nd Horse. (8) 4th Horse. (9) 1st or Royal Dragoons. (10) 2nd or Royal North British Dragoons. (11) 3rd Dragoons. (12) 4th Dragoons.

FIG. 120.

(13) Royal Irish Dragoons. (14) 6th or Inniskilling Dragoons. (15) 7th Dragoons. (16) 8th Dragoons. (17) 9th Dragoons. (18) 10th Dragoons. (19) 11th Dragoons. (20) 12th Dragoons. (21) 14th Dragoons.

A chart to show Morier's series of paintings of Horse and Dragoons, 1751, Colours in Heraldic lines—see chart p. 92.

Colour of the several Facings.	Regiments.	Clothing of the Serjeants, Corporals, and Private Men.				Clothing of the Drummers, Trumpeters, and Hautbois.		
	Ranks and Titles of the several corps of Dragoon Guards, Horse and Dragoons.	Colour of the Facings and Lapells.	Colour of the Button-holes and how the Buttons are set on.	Colour of Waist-coats and Breeches.	Hat Lace.	Cōlour, Facing and Lining of the Coats.	Colour of Waist-coats & Breech-es.	Colours of the Lace on the Clothes of the Drummers, Trumpeters, &c.
Blue	1st or King's Regiment of Dragoon Gds.	blue, with half lapells	yellow, 2 & 2	blue	gold	red with blue	blue	royal lace, yellow and blue
	1st Horse	pale blue, lapelled	white, 2 and 2	pale blue	silver	pale blue with red	red	white, with a red stripe
	1st or Royal Dragoons	blue, without lapells	yellow, 2 and 2	blue	gold	red with blue	blue	royal lace
	2nd or Royal North British Dragoons	ditto ditto	white, 2 and 2	blue	none	red with blue	blue	royal lace
	3rd or King's Own Regt. of Dragoons	light blue, ditto	yellow, 3 and 3	light blue	gold	red with blue	blue	royal lace
	5th or Royal Irish Drag'ns.	blue, ditto	white, 3 and 3	blue	silver	red with blue	blue	royal lace
Yellow	3rd Regt. of Horse, or the Carabineers	pale yellow, lapelled	white, 2 and 2	pale yellow	silver	pale yellow with red	red	white, with a red stripe
	6th, or the Inniskilling Dragoons	full yellow, without lapells	white, 2 and 2	full yellow	silver	full yellow with red	red	white, with a blue stripe
	8th Regt. of Dragoons	yellow, ditto	white, 3 and 3	yellow	silver	yellow with red	red	white, with a yellow stripe
	10th Regt. of Dragoons	deep yellow, ditto	white, 3, 4 and 5	deep yellow	silver	deep yellow, with red	red	white, with a green stripe
	14th Regt. of Dragoons	lemon colour, ditto	white, 3 and 3	lemon colour	silver	lemon colr. with red	red	white, with red & green stripe
Buff	2nd or Queen's Regt. of Dgn. Guards	buff colour, with half lapells	yellow, 3 and 3	buff colour	gold	red with blue	blue	royal lace, yellow and blue
Black	4th Regt. of Horse	black, lapelled	yellow, 2 and 2	buff colour	gold	buff colour with red	red	white, with a black stripe
	9th Regt. of Dragoons	buff colour wtht. lapells	white, 2 and 2	buff colour	silver	buff colour with red	red	white, with a blue stripe
	11th Regt. of Dragoons	buff colour, ditto	white, 3 and 3	buff colour	silver	buff colour, with red	red	white, with a green stripes
White	3rd Regt. of Dgn. Guards	white, with half lapells	yellow, 2 and 2	white	gold	white with red	red	yellow, with a red stripe
	7th or the Queen's Regt. of Dragoons	white, without lapells	white, 3 and 3	white	silver	red with blue	blue	royal lace, yellow and blue
	12th Regt. of Dragoons	ditto, ditto	white, 2 and 2	white	silver	white with red	red	yellow, with a green stripe
Green	2nd Regt. of Horse	full green, lapelled	yellow, 2 and 2	full green	gold	full green with red	red	white, with a red stripe
	4th Regt. of Dragoons	green, without lapells	white, 2 and 2	green	silver	green with red	red	white, with a blue stripe
	13th Regt. of Dragoons	light green, ditto	yellow, 3 and 3	light green	gold	light green with red	red	white, with a yellow stripe

Given at Our Court at Kensington, this 1st day of July, 1751,

OF THE SEVERAL CORPS OF CAVALRY, IN THE CLOTHING, FURNITURE, AND STANDARDS.

Housings and Holster Caps.			Standards and Guidons.				
Colour of the Housings and Holster Caps.	Colours of the Lace on the Housings and Holster Caps.	Badge or Device on the Housings and Holster Caps.	Colour of the 2nd and 3rd Standard or Guidon.	Embroidery on the Three Standards.	Fringe on the Three Standards or Guidons.	Badge or Device on the Second and Third Standard or Guidon.	Motto on the Second and Third Standard or Guidon.
red	royal lace	King's Cypher within the Garter and Crown	blue	gold	gold	King's Cypher within the Garter	
pale blue	white and red stripe	Rank of the Regiment I. H.	pale blue	gold and silver	gold and silver	Rank of the Regiment I. H.	
red	royal lace	Crest of England within the Garter	blue	gold	gold	Crest of England within the Garter	
blue	royal lace	Thistle within the circle of St. Andrew	blue	gold and silver	gold and silver	Thistle within the circle of St. Andrew	Nemo me impune lacessit.
light blue	royal lace	White Horse within the Garter	light blue	gold	gold	White Horse within the Garter	Nec aspera terrent.
blue	royal lace	Harp and Crown	blue	gold and silver	gold and silver	Harp and Crown	
pale yellow	white and red stripe	Rank of the Regt. III. H.	pale yellow	gold	gold	Rank of the Regt. III. H.	
full yellow	White and blue stripe	Castle of Inniskilling within a wreath	full yellow	silver	silver and blue	Castle of Inniskilling	
yellow	white and yellow stripe	Rank of the Regt. VIII. D.	yellow	silver	silver and yellow	Rank of the Regt. VIII. D.	
deep yellow	white and green stripe	Rank of the Regt. X. D.	deep yellow	silver	silver and green	Rank of the Regt. X. D.	
lemon colour	white, red, and green stripe	Rank of the Regt. XIV. D.	lemon colour	silver	silver and red	Rank of the Regt. XIV. D.	
buff colour	royal lace	Queen's Cypher within the Garter	buff colour	gold	gold	Queen's Cypher within the Garter	
buff colour	white and black stripe	Rank of the Regt. IV. H.	black	gold	gold and silver	Rank of the Regt. IV. H.	
buff colour	white and blue stripe	Rank of the Regt. IX. D.	buff colour	silver	silver and blue	Rank of the Regt. IX. D.	
buff colour	white and green stripe	Rank of the Regt. XI. D.	buff colour	silver	silver and green	Rank of the Regt. XI. D.	
white	yellow and red stripe	Rank of the Regt. III. D. G.	white	gold and silver	gold and silver	Rank of the Regt. III. D. G.	
white	royal lace	Queen's Cypher within the Garter	white	gold	gold	Queen's Cypher within the Garter	
white	yellow and green stripe	Rank of the Regt. XII. D.	white	silver	silver and green	Rank of the Regt. XII. D.	
full green	white and red stripe	Rank of the Regt. II. H.	full green	gold	gold	Rank of the Regt. II. H.	Vestigia nulla retrorsum.
green	white and blue stripe	Rank of the Regt. IV. D.	green	silver	silver and blue	Rank of the Regt. IV. D.	
light green	white and yellow stripe	Rank of the Regt. XIII. D.	light green	silver	silver and yellow	Rank of the Regt. XIII. D.	

in the twenty-fifth year of Our Reign.

By His Majesty's Command,

H. FOX.

CAVALRY: APPENDIX III

References to Cavalry clothing from lists, contracts, Inspection returns, etc.

1ST OR KING'S DRAGOON GUARDS

1754. Uniform lapelled with blue, blue waistcoats and breeches; yellow buttonholes and buttons; gold laced hats. Buff accoutrements coloured yellow. Officers' horse furniture red, laced gold. Drummers on greys.

1755. Officers' uniforms not according to regulation.

1755. Oct. Officers in red frocks, faced and lapelled with blue; red waistcoats laced with gold.

1756. Officers, new uniforms.

1759. Light Troop, brass helmets.

2ND, QUEEN'S DRAGOON GUARDS

1753. Officers' uniforms new and according to regulations. Buff accoutrements coloured yellow. Horses all bays, except Drummers', who ride greys. Uniform red, lapelled, faced and lined buff; yellow buttonholes; gold laced hats.

1759. Drummers and Farriers on greys. Drums brass and embossed, not painted.
Light Troop, brass helmets.

3RD, DRAGOON GUARDS

1753. Officers' and men's clothing and horse furniture good and uniform, but not according to regulations.

1755. Brass Drums. Uniform faced, lapelled and lined white, yellow buttonholes and buttons. Gold laced hats.

1759. Light Troop brass caps.

5TH, HORSE (4TH DRAGOON GUARDS)

1734. Gold shoulder knots.

1735. Horse furniture laced and fringed with gold fringe. (For officers).

6TH, HORSE (5TH DRAGOON GUARDS)

1731. Silver fringe, silver lace and silver buttons for officers.

7TH, HORSE (6TH DRAGOON GUARDS)

1732. Feathered hats for officers.

1737. Scarlet cloth, yellow ratinet linings, yellow Genoa velvet lapels, silver buttons, silver braid. Laced hats for officers. Broad silver lace for binding and tassels (for housings), silver chain lace for belts, silver shoulder knots. Waistcoats bound with silver lace.

8TH, HORSE (7TH DRAGOON GUARDS)

1729. Red cloth, black velvet, buff cloth and gold lace fringe.

1ST, ROYAL DRAGOONS

1753. Officers' uniforms embroidered with gold. Black horses. Sergeants blue sashes. Their horses have hunters' tails. Buff accoutrements coloured yellow. Horses lighter than most other Dragoons. Drummers and Farriers on greys.

3RD, KING'S OWN DRAGOONS

1752. Kettle Drums and banners received.

1755. Uniform red, faced and lined light blue, blue waistcoat and breeches, yellow buttons and holes, gold laced hats.

4TH, DRAGOONS

1754. Officers' Horse furniture laced silver. Buff accoutrements, coloured yellow. Uniform red faced and lined green; green waistcoat and breeches; white buttons and holes. Silver laced hats. Drummers on greys.

1755. Officers' uniforms proper, except waistcoats, which are buff, but will immediately have them right, viz: peagreen. Drummers are all Blacks.

1758. Officers very well mounted on bay horses. Drums of brass, embossed and not painted.

5TH, ROYAL IRISH DRAGOONS

1731. For officers: gold shoulder knots, gilt buttons; housings and holster caps laced with gold and gold tassels for those of captains. Sergeants gold shoulder knots and lace. Corporals gold lace. 96 yards gold lace and 3 yards blue velvet for Kettle Drums. 216 goat skins for holster covers.

Uniform red, faced and lined blue, buff waistcoats, red breeches, yellow buttons and holes, gold laced hats.

1754. Officers' Horse furniture not all new or alike.

1755. Brass mounted swords for Sergeants and Corporals. Officers' Horse furniture not regulation, Field officers, Captains and Subalterns not being alike.

1757. Light Troop on black horses. Uniform red faced blue, waistcoat and breeches, yellow buttons and holes, gold laced hats.

2ND OR ROYAL NORTH BRITISH DRAGOONS (SCOTS GREYS)

1754. Sergeants' horses have hunters' tails. Buff accoutrements coloured white. Uniform red, faced and lined blue, blue waistcoats and breeches, white buttons and holes.

1756. The Light Troop has 65 swivel belts, 65 cartouche boxes and belts, 63 spades, 32 handbills and 31 axes.

1759. The farrier to cock all the new hats and the old to be kept to serve as nose bags.

All foraging parties to wear frocks and foraging caps.

The men to march with their frocks over their coats.

The men are always to have the oilskin covers of their caps with them, and put them on in bad weather. Cloaks to be rolled with blue side outmost.

Lost: a dark grey horse with hunting saddle and goat skin.

1764. Waistcoats and breeches to be changed from blue to white, and buttonholes on waistcoat to be plain without lace.

6TH, INNISKILLING DRAGOONS

1739. Sergeant pouch sling and waist belt laced with gold.

1751. Uniform red, faced and lined yellow, white buttons and holes; silver laced hats.

7TH, QUEEN'S REGIMENT OF DRAGOONS

1753. Officers' uniform faced and lapelled with white, white waistcoats, red breeches, richly laced with silver. Horses of different colours. Drummers on greys. Remarkably pretty horses. Buff accoutrements coloured yellow. Uniform (men) red, faced and lined white, white waistcoats and breeches, white buttons and holes. Silver laced hats.

1758. White shag breeches.

8TH, DRAGOONS

1735. Scarlet cloth, yellow shalloon (for lining). Officers silver laced coats with silver shoulder knots, silver buttons. Sergeants silver shoulder knots, Corporals silver and yellow silk mixed shoulder knots. Blue drab cloth, blue buttons (for Drummers?) 6 Scimitars for Drummers.

10TH, DRAGOONS

1741. A portrait of John Tempest, Cornet: all scarlet coat, narrow silver loops on the chest buttons at outer end, at equal distances. The cuff has a long turn-up extending the whole length of the forearm, with 5 buttons and loops on the turn-up and loops to correspond on rear of sleeve. Silver shoulder knot over left shoulder; waistcoat appears a kind of lilac tone.

1754. Buff accoutrements coloured yellow. Uniform red, faced and lined yellow; yellow waistcoats and breeches, white buttons and holes, silver laced hats. Drummers and Farriers on greys.

11TH, DRAGOONS

1753. Brown horses.

1755. 104 brown horses. Horses not so uniform as in other regiments from scarcity of dark brown horses, on which the regiment is mounted. Uniform red, faced and lined buff, buff waistcoats and breeches, white buttons and holes.

13TH, DRAGOONS

1732. Scarlet cloth; 22½ ells of grey cloth; silver buttons and lace. Broad and narrow silver lace; silver thread buttons; silver laced hats.

1751. Officers: gold lace and wore their sashes over left shoulder.

General Forbes in his *History of the Army Ordnance Services* says that the framework of the saddle of the Middle Ages consisted of two arches joined by a single bar in the centre and one on either side from which the stirrups were suspended. These bars later developed into the "trees," and took the weight off the horse's back. General Forbes continues to say that "by the eighteenth century the saddle had become more or less standardized, underneath was the pad in one piece and next the tree. Over this was a pad of leather and fur on which the rider sat, the various parts being held together by a surcingle" (Fig. 121).

Fig. 121. Saddle, *c.* 1712.

A cloak pillion is mentioned, and was a sort of pad attached to the rear of the saddle, and on which the cloak or valise rested. This is not visible in pictures of the period. A perfect specimen of a military saddle exists, which was reproduced in the *Journal of Army Historical Research*, Vol. XVII. It is the saddle used by John Ferguson of Craigdarrock when he served as Lt.-Colonel of Viscount Kenmure's Regiment of Foot (26th) at the battle of Killiecrankie, 1689. The saddle has remained in the possession of the descendants of the original owner ever since (Fig. 122).

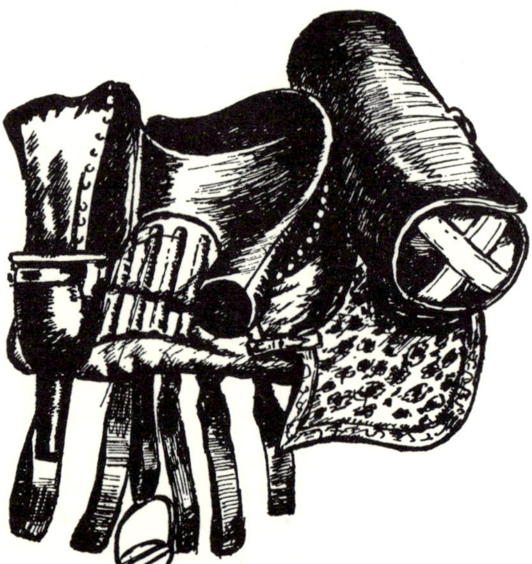

Fig. 122. Officer's Saddle, 1689.

Lt.-Colonel Wallace Cuninghame has not only allowed me to quote from his article, but has most kindly given me further information on the subject. "During the battle, Craigdarrock as was the custom served on foot, the senior Major and Adjutant remaining mounted. When the line was broken by the charge of the Highlanders his orderly mounted his charger and fled to Craigdarrock where he was cursed by the Colonel's widow, saying that his descendants should never again see a horse. Certain families in the Glencairn Valley said to be descendants of the orderly, still suffer from a form of hereditary ophthalmia."

FIG. 123. Saddle and Valise, 1689.

The saddle itself is of black leather with brass studs, mounted on a wooden tree with iron clamps. The seat is remarkably narrow, and the rider must have ridden on his fork with long stirrup leathers; these last are placed quite far back. The stirrups and buckles are of iron. Leather holsters are attached to iron D's on the tree. The valise is of leather and is fastened with an iron padlock and chain, to secure the contents from any pilfering hands. In the rear of the saddle is a leather pad or pillion, so that the valise did not rest directly on the horse's back. Strangely enough there are no D's or any means of attaching the valise to the cantle, so Colonel Cuninghame suggests it could only have been secured by a surcingle passing through the leather D's or loops on the end of the valise and going under the horse's belly, was fastened to the girth, there being an extra strap under the saddle flap, which does not appear to be required for the girth. The accompanying sketch by Colonel Cuninghame explains the idea (Fig. 123). The housings and holster caps are of leopard skin edged with silver or gold lace. The housings were fastened by straps buckled to leather D's on the saddle. The holster caps have been removed to show the holsters (Fig. 124).

FIG. 124. Holster Cap, 1689.

The bridle seems at first to have consisted of a long and a short buckling piece, both having buckles on each side, which passed through leather D's on the brow and nose bands. There was also a throat lash kept in place by a crown piece. In France a small chain was strapped over the crown piece to protect the horse's head from sword cuts. This is not seen in English pictures, although it is not impossible that it was used on service.

FIG. 125. Bridle of King William III.

In the sketch of William III's bridle it will be noticed that the throat lash and long buckling piece are cut out of one piece of leather and not being separate the crown piece is absent. There are additional ornamental straps which would not be seen on a trooper's horse.

In looking at the pictures of the end of the seventeenth and early years of the eighteenth century no sign of a bridoon appears (Fig. 126 D). In the 1742 Clothing Book a bridle similar to the one just described is shown, but with a bridoon fastened apparently to the short buckling piece of the bit. Mr. Rousselot, the well-known artist expert on military antiquities, has very kindly lent me his studies based on the sketches and descriptions in de la Porterie's *Institutions militaires*, which explains how the bridoon was fastened to the curb by a supplementary short buckling piece (Figs. 126 A and B). The bridoon was often of white raw hide, and was left loose on the horse's neck instead of being held in the bridle hand. The bit it will be noticed has one or sometimes two chain cross pieces, which served to strengthen it and keep the cheeks from being pulled out of the straight (Figs. 125, and 126 D).

On coming to the Morier paintings we find that in most cases the bridoon is fastened to separate buckling pieces (Figs. 126 C and 127). The Royal Dragoons, however, are shown with the single bit rein as previously described.

(Fig. 128 B). Head stalls of hemp, or raw hide, figure in the saddlery lists. These were used to save the proper head stall when fastening the horses in the lines. Attached to it was a length of raw hide which was carefully whitened and twisted and attached to the front of the saddle on the near side when on the march.

Watering bridles, also of raw-hide, with bridoon bits are mentioned. These, as the name indicates, were used when going to water and also, de la Porterie states, for use in training young horses, or for those whose mouths had been hurt or spoilt (Fig. 128 A).

From the presence of a carefully whitened head stall beneath the black leather one and supporting the bridoon it seems possible that it was used as

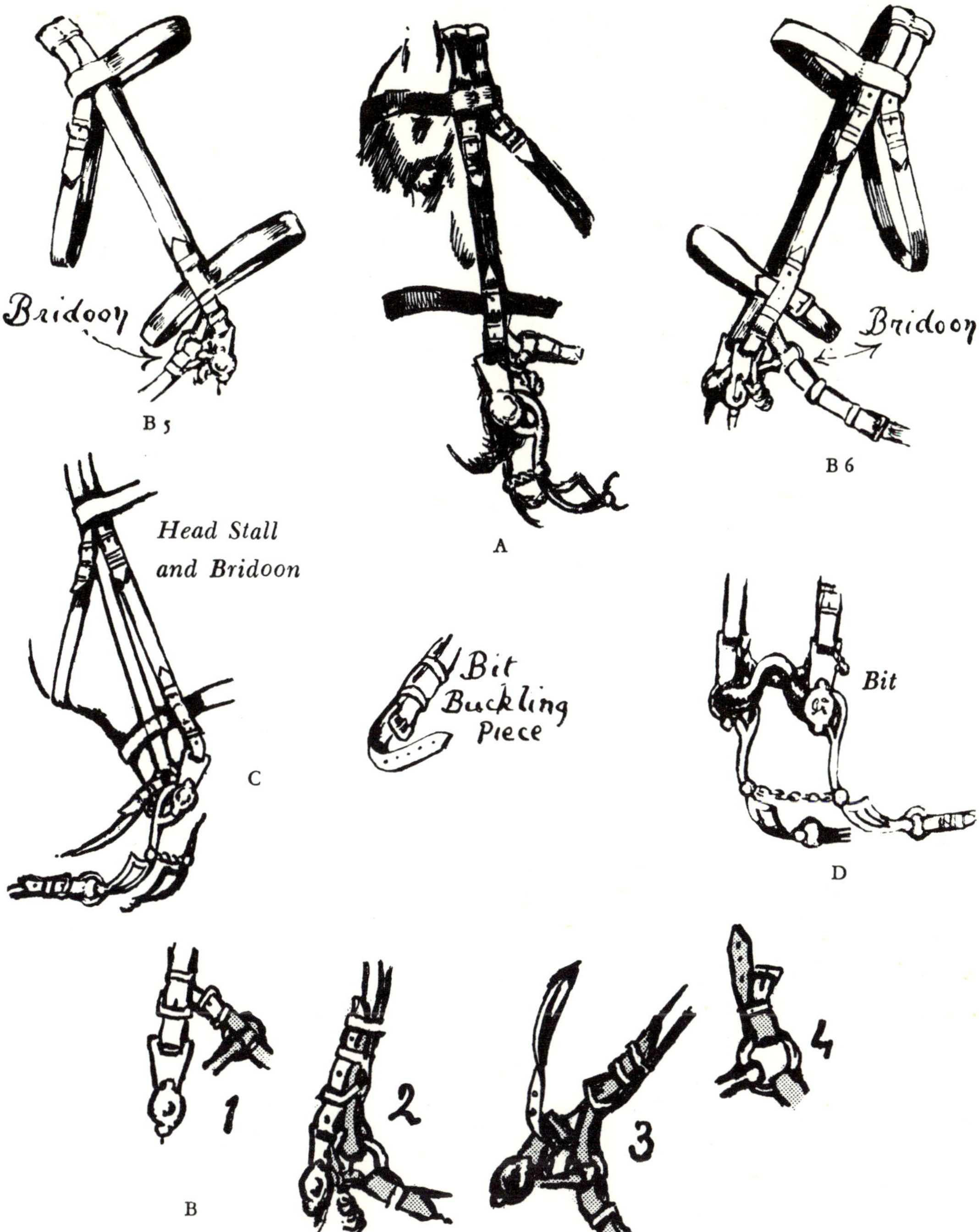

Bridoon
B 5
A
Bridoon
B 6
Head Stall
and Bridoon
C
Bit
Buckling
Piece
Bit
D
1
2
3
4
B

Fig. 126.

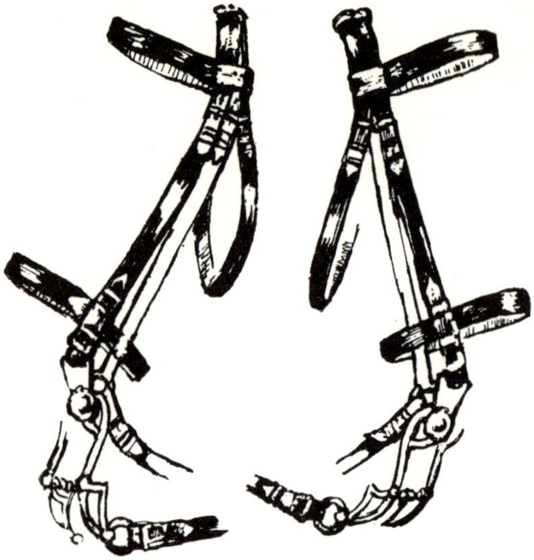

Head Collar. FIG. 127. Bit and Bridoon, 1751.

service in the dual capacity of head stall and watering bridle (Fig. 127). The bridoon rein was in one piece, whereas the curb rein was in two fastened at the end by a leather knot or "Turk's head" instead of the buckle of to-day. It was also provided with a leather runner. On looking at the Dragoon in the 1742 Clothing Book and Morier's paintings one can see the man shortening his reins by sliding the runner towards his horse's neck.

The saddle was also furnished with breast piece and crupper.

The stirrups are shown of either the square or the rounded shape. The square form is mostly represented in the Blenheim tapestries, the 1742 Clothing Book and the series of paintings, although both patterns seem to have been in use at the same time (Fig. 129 A and B).

It is curious that no valise is shown in any of the paintings of English cavalry of the date, but only the rolled cloak. It has been suggested that from

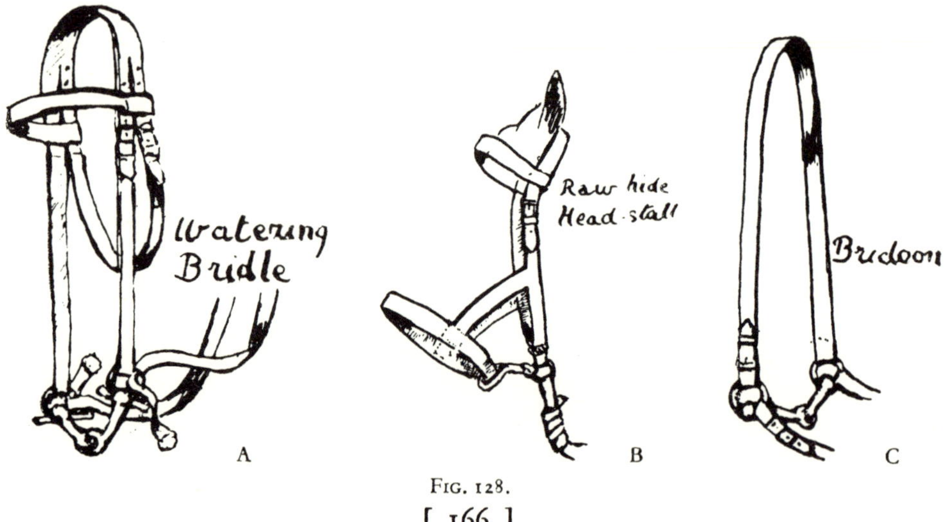

A B C

FIG. 128.

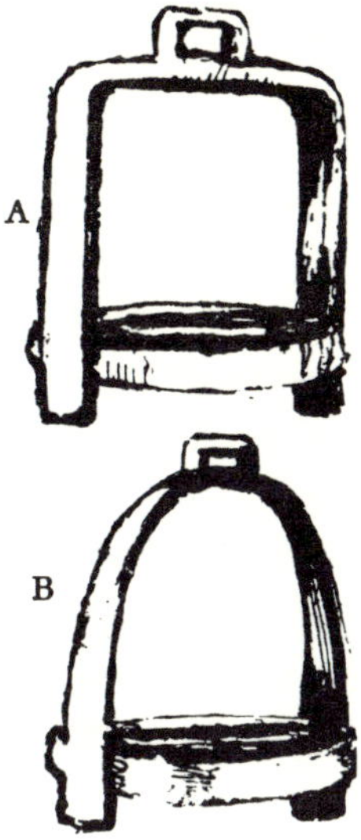

the size of the roll in some pictures various articles were packed in it, but except for items of clothing this would have been impracticable. The men, however, must have had some means of carrying their groom kit and other necessaries. In only one painting, and that of a Hanoverian, does Morier show a valise, haversack, water-bottle and feed bag. The valise is the round type of red cloth, the water-bottle is attached by a thick cord and is the flat wooden barrel shape of the 1800–15 period. The feed bag is just like the modern article. In addition the man carried his tent pole fastened to his musket (Fig. 130). Spare horse-shoes are not visible in any picture; these in the French Cavalry were carried strapped to the rear of the saddle on either side in much the same position as in modern times.

FIG. 129. Stirrups.

At the proceedings of the Court Martial of Lord George Sackville, it appears that it was customary for the Cavalry in preparing to charge "to throw away their picket poles and everything that encumbered them." Colonel Sloper in his evidence stated that "he was not ready to charge as he had not ordered his men to throw away their picket poles."

FIG. 130. Trooper's Marching Order, *c.* 1751.

1714–1760

FROM 1714 to 1743 there are one or two examples of colours, the first being those of Grove's regiment, 10th Foot, which are reproduced by Milne and are taken from a MS work entitled *Regimental Book of Records*, 1726. This book is embellished with chapter headings of trophies of the regiments' "colours." One of these is the union throughout without any device. That described by Milne as the Colonel's is white with heraldic ermine spots in the centre a black talbot on a red and white wreath over which is a yellow label with the motto *Vinco Malum Bono*. These were all parts of the Colonel's coat of arms. In the upper canton is a small union. Two small yellow camp colours with black talbots are also depicted. Behind these is another pair of colours and two more camp colours forming part of the trophy; these all have the Sun in splendour with a face in the centre, possibly colours carried at an earlier date. Grove became Colonel in 1715, so no doubt these colours were taken into use soon after that date.*

FIG. 131. Colour and Piled Halberds.

Another MS work similar to that of the Trophies of Louis XIV shows those of Louis XV and the colours captured by his armies. Amongst them is a pair of colours of the 23rd or Royal Welch Fusiliers taken at Fontenoy 1745. These have a small union in the upper canton with the white horse in the centre of the St. George's cross; in the four other corners in cartouches are the red dragon, the rising sun and the three feathers; in the centre are the Prince of Wales's feathers and motto *Ich Dien* surrounded by branches of laurels in proper colours and tied with pink ribbons, the whole on a white flag. The second colour is identical except that the devices are on a crimson colour and the stalks and branches are painted green instead of brown and the ribbon is white.

* Reference: "Colours and Standards," by S. M. Milne.

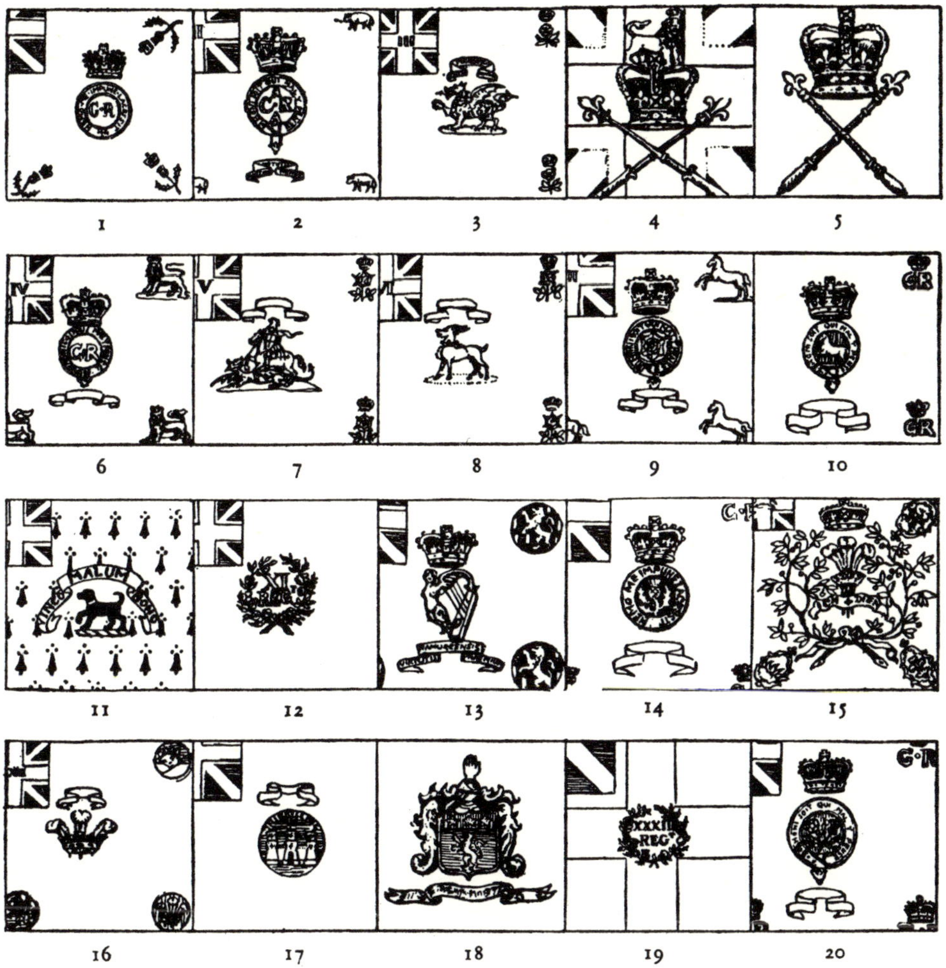

FIG. 132. Colours of Regiments of Foot, 1715–1751.

(1) 1st or Royal Regiment. (2) 2nd or Queen's. (3) 3rd or The Buffs. (4 and 5) 4th or King's Own, 1745. (6) 4th or King's Own, 1751. (7) 5th, (8) 6th, (9) 7th The Royal Fusiliers. (10) 8th or King's. (11) Grove's or 10th, *c.* 1725. (12) 11th, 1751. (13) 18th or Royal Regiment of Ireland. (14) 21st Royal North British Fusiliers. (15) 23rd or Royal Welch Fusiliers, 1742. (16) 23rd or Royal Welch Fusiliers, 1751. (17) 27th or Inniskilling. (18) Handasyd's Regiment, 1742. (19) 32nd, (20) 21st or Royal North British Fusiliers.

Another colour captured at Fontenoy is that of Brigadier-General William Handasyd's regiment, the 31st. This appears to be white with the Colonel's arms in the centre. In neither of these cases do the colours match the facings of the regiment, those of the 23rd having blue and those of the 31st being buff, unless in the latter case the colour was originally a pale buff and has faded with time.

The beautiful pair of colours with their rich design of the 4th King's Own Regiment are of special interest, not only for their antiquity but also for their romantic history. The crown and crossed sceptres are very different from the design of the 1751 colours, and one would like to know when they were first emblazoned on the Regiment's colours. These colours were carried by the Regiment at the Battle of Culloden (and no doubt previously).

Together with the Stewart of Appin clan flag they were gifted to the Scottish National Naval and Military Museum, Edinburgh, by Mr. Stewart of Ballachulish, with the proviso that "as they had opposed each other on the field of battle, so now shall they ever rest together in peace".

We now approach the important period when the "colours" of regiments, like their uniforms, were to conform to certain definite rules, which first appear in the drafts of 1743 and 1747 (already referred to) and which culminated in the Royal Warrant of 1751.

Possibly because the details of "colours" in the draft were not sufficiently clear, regulations were issued in the General Officers Letter Books entitled "Regulations for the colours of the Marching regiments 1747 as delivered to the Clothing Board November 11th, 1749. Robert Napier Adjutant-General."

At the same time Napier had a volume of coloured drawings made, showing the various ensigns, standards and guidons and depicting the badges and the manner and position they were to be placed on the flags. In addition there are diagrams giving their dimensions and measurements. This invaluable book is now in the Royal Library, Windsor.

By the Warrant of 1751 the Colonels' coats of arms or badges, which had previously formed a very prominent feature of the decoration of the regimental colours, were now abolished.

The Colonels', Lt.-Colonels' and Majors' colours which had been the only survivors of the numerous company ensigns were to be done away with and only two to be carried by Foot regiments. The first, or King's colour as it was soon to be called, was to be the great union. The 2nd, or regimental colour, was to be of the same colour as the regiment's facings with the union in the upper canton except in the case of those regiments whose facings were either red or white, when the ensign was to be white with a red cross of St. George over its entire surface with a small union in the upper canton. In the centre of both colours was to be the number of the regiment in Roman figures painted in gold within a wreath of roses and thistles on one stalk. Those regiments which had the privilege of bearing Royal devices, or ancient badges, were to have the number of the regiment painted towards the upper corner. These are mentioned as being all the Royal regiments—the (3rd) old Buffs, 5th, 6th, 8th or King's regiment and 27th, the Inniskillings. Besides these the colours of one of the marine regiments are shown bearing a full rigged man of war in the centre of both colours, the regimental one being yellow.

The mention of the ancient badges is tantalizing, there being no previous reference to them or the reason and date of their being granted. In some cases such as the 18th, 23rd and 27th the association is obvious.

Although various suggestions have been advanced as to the origin of these ancient badges, it seems clear that (whatever the reason they were granted) they were mostly Royal badges and as such are to be seen as Company badges on the colours of the Foot Guards; for example the Welch dragon of the Buffs (a badge of Henry VII); the St. George of the 5th; the antelope of the 6th (a badge of Henry IV); the Tudor rose of the 7th (a badge of Henry VIII), etc.

The badge is placed in the centre of both the King's and the 2nd or regimental colour, the numeral in both cases appearing in the upper canton. Most of the badges have the crown above except the 3rd, 5th, 6th, 23rd, and 27th, where it is absent. The label in every case is white, but often is without the motto, nor is it given in the Warrant. In the case of the 4th and 23rd Foot no label is placed on the King's colour.

The corner devices do not appear on the King's colour.

A brief description of the colouring will help to understand the chart; in every case the 2nd or regimental colour is indicated:

1ST OR ROYAL REGIMENT

Blue, with G R in gold on blue field within the green circle of St. Andrew. In the three corners a crown and thistle proper. The colours of the 2nd battalion are distinguished by a flaming ray of gold, as the pile wavy is here called; scroll without motto.

2ND OR THE QUEEN'S OWN ROYAL REGIMENT

Sea green, with CARA in gold on red within the garter, in the three corners the white lamb, scroll with motto *Pristinæ Virtutis Memor*. The letters C R stand for Carolina Regina. There are several suggestions as to the two A's; one that they are the terminals of the two previous words; another that A stands for Anspach, Caroline being the daughter of the Margrave of Brandenburgh-Anspach, and lastly that they stood for Anne the last Queen.

3RD OR THE BUFFS

Buff, with a green dragon above the scroll with *Veteri Frondescet Honore*. In the three corners a crown and rose proper.

4TH, OR KING'S OWN ROYAL REGIMENT

Blue, G R in gold on red field within the garter and plain white scroll below, in the three corners the crowned lion of England in gold.

5TH REGIMENT

Pale yellow. St. George in white armour on a brown horse, and green dragon. In the three corners a crown and rose proper. The facings are given as gosling green but unless the tone has changed with time a yellow colour was meant.

6TH REGIMENT

A deep yellow; in the centre a white antelope on a green mount, plain scroll, a crimson rose crowned in the three corners. The antelope, as in all the early representations, is more like a goat.

7TH OR THE ROYAL ENGLISH FUZILIERS

Blue, a red rose in red field within the garter. Plain scroll. In the three corners the white horse.

8TH OR THE KING'S REGIMENT

Blue. The white horse on a red field within the garter, scroll with motto *Nec aspera Terrent*. In the three corners G R and crown in gold.

18TH OR THE ROYAL IRISH

Blue. A gold harp. Motto *Virtutis Namurcensis Præmium.* In the three corners the golden lion, of Nassau in a blue field, surrounded by golden billets.

21ST OR ROYAL NORTH BRITISH FUZILIERS

Blue. The thistle on a blue ground within the green circle of St. Andrew. In the three corners G R and crown. Plain scroll.

23RD OR THE ROYAL WELCH FUZILIERS

Blue. The three feathers issuing from the Prince's coronet. Scroll with motto *Ich Dien.* In the upper corner a gold sun rising behind a green hill, blue sky. A red dragon on blue field and the feathered coronet on red in the lower corners.

27TH OR THE INNISKILLING REGIMENT

Buff. The Castle of Inniskilling white, in a blue field. Plain scroll.

41ST OR THE INVALIDS

Blue. A rose and thistle proper in a red field within the garter. Plain label. G R and crown in the three corners.

Although the 1743 Warrant had completely revolutionized the colours for the line regiments, the Guards still followed the rules originally laid down by Venn in 1672, the first colour in each regiment remaining plain crimson throughout.

THE 1ST FOOT GUARDS

The Royal standard carried by the King's Company at State ceremonies was crimson with the crown and double cypher in the centre and in the four corners the rose, thistle, fleur de lys and the harp, all with crowns over them. The Colonel's crimson with the crown in the centre; Lt.-Colonel's (King's colour of the 2nd Battalion) the union throughout with the crown and double cypher in the centre. The Major's (King's colour of the 3rd Battalion) the same as the Lt.-Colonel's but with a gold blaze or pile wavy from the upper corner. 1st Captain's the union with the crest of England crowned with the numeral I in the upper canton and so on with the remaining companies with their respective badges and numerals, being thus much the same as when the badges were first granted by Charles II. Shortly after this the Lt.-Colonel's and Major's were changed to plain crimson colours with a small union in the upper canton, the Major's having besides the small pile wavy issuing from the lower corner of the union.

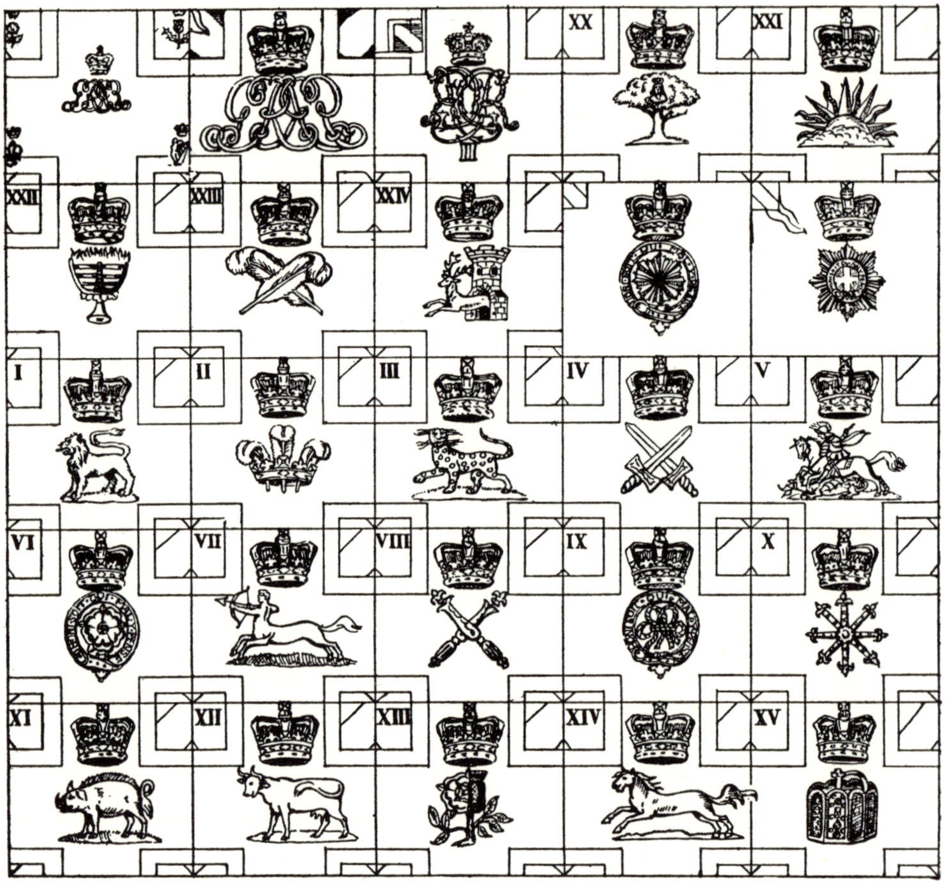

FIG. 133. Colours of 1st and 2nd Foot Guards, *c.* 1750.

TOP LEFT TO RIGHT: 1st Foot Guards. (1) Royal Standard. (2) Major's (King's colour, 3rd Battalion). (3) King's Colour, 2nd Battalion. (4) (5) (6) (7) (8) Company colours. (9) Coldstream Guards, Lt.-Colonel's colour. (10) Major's (King's colour, 3rd Battalion), the remaining numbers being Company colours.

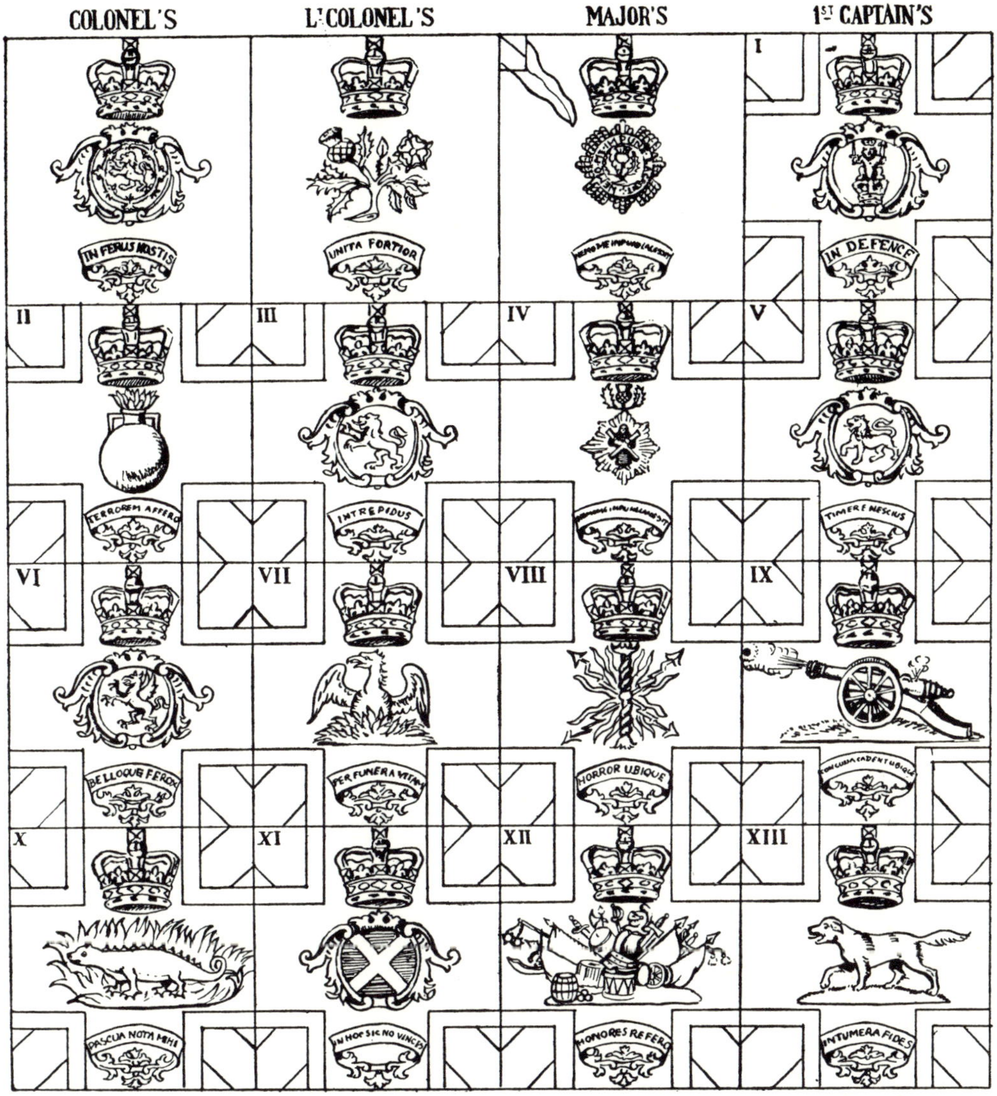

FIG. 134. Colours of 3rd (Scots) Foot Guards.

Milne shows the King's colour of the 2nd Battalion (1751–58) with the union in the upper canton, the double cypher and crown with the numeral II below (Fig. 133).

THE COLDSTREAM OR 2ND FOOT GUARDS

The Colonel's crimson with crown, garter and star, the Lt.-Colonel's the star within the garter and crown above and a small union in the upper canton. The Major's as the Colonel's but with a small union and gold pile wavy. Then follow the Company Captain's colours with their crowns, badges and the numeral in the upper corner (Fig. 133).

THE 3RD FOOT GUARDS

The Colonel's with arms of Scotland on a gold cartouche with the motto *En Ferus Hostis* on a white label, the Lt.-Colonel's crimson with the union in the upper canton and the thistle and rose on one stalk with the crown over. The Major's as the Lt.-Colonel's but with the pile wavy from the lower corner of the union. In the centre of the colour the star of the Order of the Thistle with crown over. The Company Captains' colours were unions throughout with their crowned badges and numerals in the upper canton.

It will be noticed that there are slight differences from the early list, the order of the Company badges being changed. The 14th, 15th and 16th Companies are shown as the 2nd, 3rd and 4th, and the Lion passant becomes the 5th. The Griffin rampant azure now vert is numbered the 6th. The motto *Concussæ* becomes *Concuss*, and that of *Terroram* is now *Terrorem affero; Intumera Fide* becomes *Intumera*. The two colours with the same device, viz: Lion rampant and St. Andrew's cross become Lion rampant and burning grenade (Fig. 134).

In all the regiments of Foot Guards one only of the Company colours was used in rotation as the regimental colour of a battalion.

The long series of the badges of the various Sovereigns granted to the Foot Guards and still emblazoned on their colours in much the same way is a tradition not to be found in any other country and who, on glancing at the records of these regiments through nearly three centuries, can but say how worthily they have upheld the honour of their Royal privileges?

CAVALRY STANDARDS

1714–1760

PERHAPS the only cavalry standard existing of this period is preserved in the Royal United Service Museum. It is that of the 8th Horse (7th Dragoon Guards). This, according to Milne, was the standard in defence of which Cornet Richardson received upwards of thirty sabre cuts at Dettingen. It was presented to him after the action and was preserved in the family until it found its present home. It is of crimson damask with a gold fringe. The obverse side is embroidered with the crest of General Ligonier, a demi lion issuing from a ducal coronet and his motto *Quo Fata Vocant*. On the other side is his full coat of arms surrounded with a trophy of arms, and what may possibly be the regiment's camp colours, the whole richly worked in gold. A small union about 3 inches square occupies the upper canton. Milne puts its date as about 1720 when Ligonier became Colonel of the regiment.

The only other representations of standards, or rather guidons, are those shown in the book of trophies of Louis XV, which depicts amongst those captured at Fontenoy 1745 a yellow guidon with the arms and crest of Robert Rich and his wife on one side and a double G R on the other.

Another Dragoon guidon, which has not been identified, is one of green damask with a gold design worked round the border, the fringe being silver. In the centre is a stag proper on a green mount, the background and cartouche both being silver, below on a white ribbon is the motto *Rouze*. A small union is in the upper canton. According to Bland in his *Treatise on Military Discipline*, 2nd edition, 1727, the cavalry now only had one standard per squadron, so probably the Troop badges of such regiments as the Horse Guards and Royal Dragoons had now disappeared.

The 1743–51 Warrants ordered that the standards and guidons of Dragoon Guards and standards of Horse were to be of damask embroidered and fringed with gold or silver. The guidons of Dragoons to be of silk, tassels and cords of the whole to be of crimson silk and gold mixed.

The King's or First standard or guidon of each regiment to be crimson with the rose and thistle conjoined and crown over in the centre; His Majesty's motto *Dieu et mon droit* underneath; the white horse in a compartment in first and fourth corners, and rank of regiment in gold or silver characters on a ground of the same colour as the facing of the regiment in a compartment in the second and third corners.

FIG. 135.
(1) Household Cavalry. (2) Royal Horse Guards. (3) 2nd Dragoon Guards. (4) 8th Horse.
(5) Unidentified Dragoon Regiment, 1742. (6) Rich's Dragoon, 1742.

The Second or Third standard, or guidon, of each corps to be the colour of the facing with the badge of the regiment in the centre, or the rank of the regiment in gold or silver Roman characters on a crimson ground within a wreath of roses and thistles on the same stalk, the motto of the regiment underneath. The white horse on a red ground to be in the first and fourth compartments and the rose and thistle conjoined upon a red ground in the second and third compartments, the distinction of the third standard, or guidon, to be the figure 3 on a circular ground of red underneath the motto. Those Corps which have any particular badge are to carry it in the centre of their second or third standards, or guidons, with the rank of the regiment on a red ground within a small wreath of roses, thistles in the second and third corners.

The 1751 Warrant gave full details of colour, fringe, devices, mottoes, etc. for each regiment of Life Guards, the 1st and 2nd Horse Grenadiers, the Horse Guards and the three guidons of the 2nd Dragoon Guards.

The Windsor Colour Book only gives the following standards and guidons:

1st TROOP OF LIFE GUARDS

A standard and a guidon of crimson damask fringed with gold and both embroidered with the same devices, in the centre the rose and thistle on one stalk proper, on a gold ribbon the motto *Dieu et mon Droit* with three small crowns underneath, the whole surmounted by a large crown with G R on either side. The Second Troop had standard and guidon embroidered in the same way but of white damask.

The 3rd and 4th Troops had been disbanded in 1746 so their standards are not shown.

1st TROOP OF HORSE GRENADIER GUARDS

A square standard and a guidon embroidered as above but of blue damask and the motto ribbon is silver.

2nd TROOP OF HORSE GRENADIERS

As the first but differing only in being of crimson damask.

THE ROYAL HORSE GUARDS BLEW

1st standard, crimson damask with the Royal arms and supporters with motto on gold ribbon below and the crown above the G and R either side.

2nd standard, the Union badge of rose and thistle proper with crown over between G and R.

3rd standard, in the centre the double cypher interlaced and crown and G R as the 2nd.

All three standards are square, and of crimson damask fringed with gold and gold tassels and strings.

The three guidons of the 2nd Dragoon Guards are the same shape as those of the Horse Grenadiers but have the same devices as those of the Royal Horse Guards except that the 1st is crimson, the 2nd blue, and the 3rd buff, and are fringed with gold, and the colour of the individual guidon.

FIG. 136. Train of Artillery, 1722.

ARTILLERY AND ENGINEERS

1714–1760

AFTER the shameful Peace of Utrecht by which the politicians, for party gains, threw away the successful results of ten years of war, the Train of Artillery which had formed part of Marlborough's ever triumphant army was disbanded according to the usual practice.

The campaign in Scotland against the Old Pretender proved the utter futility of the system, and it was obvious that a permanent corps of artillery was a vital necessity.

Two companies were accordingly raised by a Royal Warrant, May 1716, and we may well believe that this was largely due to the advice of Marlborough, who was Master General of the Ordnance at the time.

These Companies were formed at Woolwich, and each consisted of 5 Officers, 9 N.C.O.s, 30 Gunners, and 50 Matrosses. The first Colonel was Albert Borgard, a Dane, who had seen service on the Continent before coming to England in 1692. He had served as Second Engineer Officer of the Train in 1697 and in 1712 was appointed Chief Fire Maker. He was succeeded by Colonel Jonas Watson, who had served with distinction through the War of the Spanish Succession and remained in the Service until the age of 70, when he was killed in action while commanding the Artillery in the West Indies.

It seems most probable that it was at this time that the blue uniform faced with red was adopted, but unfortunately no definite contemporary evidence has so far come to light. Those clothing lists which have been found give a certain amount of details, as for example that of February 20th, 1715.

Waistcoats and breeches according to pattern for Sergeants, Corporals, Gunners, Matrosses and Drummers: Blew Kersey surtout coats.

January, 1726.

Allowance for "complete suits—
Blew surtout-coats 22.
Gold laced coats for 2 Sergeants and 20 Gunners and Matrosses."

These surtouts, it is true, appear to have been the undress and not the full regimentals, but whereas previously they are always described as red, like the full dress, there is now no further mention of their being any longer of this colour.

The lists also mention blew breeches and white and grey stockings, agreeing very much with the dress orders of 1742.

The print of the funeral procession of the Duke of Marlborough does not show the Gunners dressed in red, like the other troops, but in coats of a faded greenish tone, which originally may have been blue. They wear Grenadier caps with hanging hoods. The drivers wear cocked hats and long coats with stitched button holes which may have been blue or grey, but the colours are too faded to say anything definitely, except that they were not red (Fig. 136).

The Ordnance Board, like the Admiralty, had its own Officers' barge to convey the members up and down the Thames to the Tower, the crew of watermen being dressed in blue lined with blue blaize; that of the Admiralty, curiously enough, wearing red coats. The crew would no doubt have worn a metal badge on the arm, the customary insignia of a waterman.

The equipment lists give the following arms and accoutrements:

1715. The list of stores for the Train, besides halberts and drums, has the following:

Quarter colours on staves shod with iron.

Ensign 8 yrds fly and 8 breadth with 4 unions in one quarter.

White pendant, 15 yds. fly 1 yd. broad, at the head, with three black balls.

Jack flags and Union flags of beau port with hallyard.

1716. Hangers with ordinary hilts, and new scabbards, hilts and handles black, hangers with iron handles and brass hilts, snaphance muskets, musketoons, muskets, pistols, buckled shoes and leather garters.

1728. Slings for Gunners, buck slings for (powder) horns and muskets, stitched (i.e. laced) sword belts for Sergeants, Corporals, and Bombardiers. For the Gunners buck slings for powder horns, and sword belts unstitched.

FIG. 137. Officer of Artillery, 1742.

For a Matross a musket sling instead of one for a powder horn. For a Drummer a cloth belt and sling. For the expedition to Vigo, 1719, ten watermen and a coxswain were included in the personnel of the Train. (Cleveland Notes.)

The copy of the 1742 Clothing Book, which is in the library of the Royal Artillery Institution at Woolwich, has two highly finished original water colours. These were no doubt made to complete the series of uniform plates, but were never printed. They represent an officer and a Gunner of the Artillery and agree with the orders for dress of 1743 (Figs. 137 and 138).

Officers—a blue double-breasted coat with blue lapels and brass buttons, red cuffs, blue stitched buttonholes, red skirt linings, red breeches and waistcoat, the latter laced on the edges with gold; gold laced hat and white spatterdashes; a crimson sash worn over the left shoulder, white buff waistbelt and cartouche box, which is ornamented with a brass crown on red, probably a piece of red cloth; long brass hilted sword and a fuzil, but no bayonet (Fig. 137).

The Gunners and Matrosses plain blue coats, all N.C.O.s and privates having scarlet half lapels, scarlet cuffs and slashed sleeves with 5 buttons, blue waistcoats and breeches; the Sergeants' hats having a broad gold lace edging, those of the other N.C.O.s and privates a narrow one; white spatterdashes for all N.C.O.s and privates.

The Sergeants, blue coats, the lapels, cuffs and pockets of which were trimmed with a broad single gold lace.

A coarse blue surtout was used for laboratory work and fatigues.

The Sergeants, Corporals and Bombardiers were armed with halberds and brass-hilted swords. The Gunners carried field staves or linstocks, 2 feet longer than the halberts. These were kept very bright; a buff belt over the left shoulder carried a large powder horn for priming. The Matrosses were armed with muskets and bayonets.

1743. 18th May. For the use of 6 cadets and some miners ordered to Germany—

Blue cloth coats lined with scarlet shalloon laced with gold lace, scarlet cloth waistcoats lined scarlet shalloon, scarlet cloth breeches, gold laced hats, buff stitched belts and slings. From the numbers these were for ten cadets.

For the Gunners (? miners) surtout coats, hats, yarn stockings, shoes and buckles, linen spatterdashes with buckled garters and plain buff belts. The miners were attached to the artillery.

1747. March 2nd. Men were not to go to work in their regimentals, but either in frocks or surtouts.

1749. The Guard to mount in black spatterdashes and the officers in boots.

Before taking the field in this year the Gunners, field staves, powder horns and slings and swords and the muskets of the Matrosses were laid aside and both these ranks were armed with carbines and bayonets. All the N.C.O.s continued to carry halberts.

1749. March 27. Garrison order by Colonel Belford "Captains to provide their men with

FIG. 138. Gunner, 1742.

one pair of white stockings each, in which they are always to parade when not under arms."

1749. 3 April. N.C.O.s and men to be provided with 1 pair of good white leggings, with black buttons, and 1 pair of black, and a rose in their hair, the same as we had in Flanders. (The ornamental tie for their hair.)

In 1750 the buttonholes of the Sergeants' coats were to be laced with gold loops, those of the other N.C.O.s with yellow worsted; Corporals and Bombardiers wore gold and worsted shoulder knots.

Surtouts were laid aside and complete suits of clothing issued every year.

From this time onwards Sergeants wore two gold lace shoulder knots, a Corporal two of worsted, and a Bombardier one.

Garrison orders of Colonel Belford—"Officers are to mount guard in their regimentals, with fusees and cartouche boxes."

1750. March 30. The Sergeant of the main guard is not to allow any N.C.O. or private man to go out of the Warren Gate unless dressed clean, hair combed and tied up, with clean stockings and shoes well blacked.

1751. April 5th. Garrison order of Colonel Belford—"Officers on duty to wear the knots of their sashes in the left pocket, with the fringe hanging out."

1754. Jan. 2nd. "No officer to appear under arms wearing a bob wig."

1754. Halberts of Corporals and Bombardiers abolished and to be armed with carbines.

1756. Feb. 13. Captains were ordered to provide their companies with a knapsack and haversack per man. Men were on all occasions to wear their hair clubbed. In the same year a company of miners formed for service in Minorca was on its return to Woolwich incorporated in the Royal Artillery.

In 1758 officers wore boots and the men black spatterdashes on all ordinary parades, the white spatterdashes being worn on full dress occasions.

In General Orders of May 4, 1758, officers' waistcoats for their regimental frocks are to be plain scarlet, without lace, and in 1759 they discontinued carrying fuzils.

In 1760 the Royal Irish Artillery was formed as a separate regiment, the uniform at the time being much the same as that of the English. Later some minor details of dress were introduced by which it could be identified.

The portrait of Major-General Philips shows him wearing a blue coat with a laced collar, red lapels laced on the edges and buttonholes with gold, a red waistcoat with edges, buttonholes and pockets laced gold, a gold aiguilette, red breeches. A portrait of a younger officer shows an almost identical kit with the exception of the cuffs, which are laced in a slightly different way; they are red, edged gold, with a blue slash with a double edging of gold and gold loops, red breeches, a waistcoat, the latter being laced like that of General Philips (Figs. 139 and 140).

FIG. 139. Officer of Artillery, *c.* 1759.

Macdonald in his *History of the Dress of the Royal Artillery* has a picture based on the portraits of Colonels Griffiths Williams and Thomas Hosmer *c.* 1760 and Captain Tiffen, who served at Minden. This figure wears a dress similar to the above, but with a turn-down red collar and without any gold lace on the buttonholes, the cuffs, lapels or waistcoat. He wears a sash over the right shoulder, a gilt gorget and top boots.

The technical training now required by artillery officers led to the establishment by Royal Warrant in 1741 of the Royal Military Academy of Woolwich, being the first military educational institution for the training of cadets destined for a military career.

The cadets, who in 1744 numbered forty, were formed into a company which was thereafter to rank as the Senior company of the regiment. At first they wore no uniform, but later were to be dressed in the blue and red of the corps.

Soon after the raising of the regiment Woolwich became the site selected for the establishment of the arsenal as well as the depot for all Ordnance stores.

The regiment, which in 1716 consisted of two marching companies, had been gradually increased to 24 companies in 1757 divided into two battalions. The companies, however, were known by the names of their Captains.

It was no doubt due to the training at the Royal Military Academy that the English artillery was soon to establish a high reputation for the rapidity and accuracy of its fire. At Minden the guns of Philips, Foy and Macbean, to-

FIG. 140. Officer of Artillery, 1759.

gether with the infantry, were the chief contributors to the victory. Macbean with his brigade, as it was then termed, of ten 12 prs. "by the rapidity and accuracy of his fire, soon reduced the enemy's 30 guns to silence." Nor was this all, for in spite of the heavy work they had already undergone Foy and Macbean at the final stage of the action limbered up their guns and advanced with the greatest rapidity to follow the retreating masses of the enemy, which were heavily punished. For their part in the victory the English artillery were specially mentioned in Prince Ferdinand's General Order.

At Warburg, when Granby's cavalry were ordered to push forward at a trot, accompanied by the artillery, these somehow managed to keep up with them at a pace which astonished the allied Generals, who witnessed it. Again, as at Minden, on the retreat of the enemy the guns were brought forward at a gallop to pursue them with a harassing fire. It is stated that this was the first instance of guns being brought into action at a gallop.

The guns and various vehicles of the train had much the same appearance as hitherto, except that the long and comparatively straight trail of the gun carriage had now been replaced by one of a shorter and more angular shape. This form had already been used for the howitzer field carriages during Marlborough's campaigns, as it had been found from experience that the long and relatively straight trail had been liable to break going over rough ground. A foreign writer in 1750 states that "the English artillery was distinguished by its lightness, elegance, and the good quality of its materials."

During the Seven Years' War the English guns were noted for being kept the cleanest and in the best order of all the artillery in Ferdinand's army.

The lighter pieces such as 3 prs. and 6 prs. were still attached to infantry, but were served by Gunners of the artillery. Two guns were allowed to a battalion and the detachment consisted of an officer, 2 N.C.O.s and 12 men.

It is said that the battalion guns of the Guards at Fontenoy were entirely worked by crews of guardsmen, and were moved by means of drag ropes.

The 3 prs. were often mounted on the two-wheeled galloper carriage without any limber or fore-carriage, drawn by two or more horses, the wheeler being harnessed to shafts attached to the trail. It may be possible that the guns which caused such astonishment at Warburg by being advanced at the gallop were mounted on carriages of this description.

6 prs. were drawn by five or six horses and the same system of harnessing tandem-wise for the smaller pieces seems to have been still in vogue. The drivers were still civilians, ever ready to withdraw themselves and their horses out of range. It was the bolting of the drivers with their teams which resulted in the absence of guns on the field of Fontenoy with the exception of the few light battalion guns.

The evidence of the artillery officers at the Sackville court-martial gives further interesting details about our guns at Minden. It is stated that the guns consisted of medium 12-pounders, which were fired at the French cavalry at a range of 900 to 1000 yards. The officers said in their evidence that "when we come into action we take the fore wheels off" (i.e. unhook or unlimber from the fore-carriage). "We never march with more than one gun in front; as soon as we come in sight of the enemy, we draw all in a line." Usually five horses were harnessed to a gun, but at Minden a team consisted of seven. These were probably in pairs except the wheeler in single harness in the shafts.

From the following notice a kind of uniform was given to the Drivers: "The drivers attached to the train of artillery wear white frocks with letters G R in red on their backs." (The *Ipswich Journal*, 20th March, 1756.)

Guns had no sights until the beginning of the nineteenth century and were laid by a line of metal notches cut in the muzzle and breech and elevated or depressed by placing a wedge-shaped wooden block under the breech, the elevation being judged by eye.*

Morier in one of his paintings *c.* 1760 shows a gun carriage and limber. These are painted a grey colour very similar to that used in the 'eighties.

* Reference: History of the Army Ordnance Services. General Forbes.

The Kettle Drummer with his drum carriage was still with the train. His dress and that of the driver in 1742 was still the richly laced coat and hat, the details of which were much the same as before.

The Kettle Drum carriage accompanied the artillery to Vigo in 1719 and was with the army at Fontenoy, its last appearance in the field being in 1748. It seems to have remained on the establishment as late as 1756. It was considered part of the heavy baggage, and when this was ordered away the Kettle Drums accompanied it. (Cleveland Notes.)

Another curious feature of the artillery was the "flag gun". There has been much discussion concerning it, and its actual significance. From the many references to it there is no doubt that it was actually used. Cleveland says that the flag gun carried the Royal Standard and indicated the Head-quarters in the field. It seems, however, that the Royal Standard would only have been used when the Sovereign was present, as in the case of George II in Flanders. Cleveland goes on to state that "in 1757 a quater guard of the regiment of artillery commanded by a subaltern was first mounted and encamped opposite the flag gun and Kettle Drums."

The following orders also contain further references to this gun:

Viller's Camp, Aug. 7th, 1746, "a court martial to be held to-morrow morning at 9 to try all prisoners at the flag gun."

"A picquet post to be put up to-morrow morning early in front of the flag gun."

Montagny Camp, Aug. 17th, 1746. "The Captain that commands the reserve guns, to see that the waggons march with him, with the ammunition to compleat the guns to 120 rounds, and that the Gunner at the flag gun marches with a lighted match, and the rest of the Gunners to have their linstocks primed, and ready to light."

General Stringer Lawrence in India said that "his post would be known by the English colours carried on his flag gun."

A rough woodcut in the British Museum dated *c.* 1722 shows a gun and team on the march, with a Union flag attached to the gun carriage by the barrel of the piece. Another print also exists showing the flag fixed to the gun carriage. Smith's *Military Dictionary* mentions it in the long list showing the composition of a train of artillery (Fig. 137).

It seems from the foregoing that the flag gun was used to indicate the headquarters of the artillery, and also the position of those of the General commanding in the field.

For the Company in Flanders, 1749, flags with staves were used to distinguish the different sections of the train as follows: The union for the guns, red for the waggons and white for the Pontoons. (Records, W.O. 55/352.)

An early 18th century engraving of General Wade's forces in Scotland shows a party of men marking roads (Fig. 141). The uniformity of their dress makes one think that this is a party of pioneers and not hired civilian labourers. The figures are very small and minor details not clear in consequence, but they seem to have coats with stitched or lace buttonholes. It is, of course, possible that this is an infantry fatigue party, but the fact that some of the hats have their brims down and others looped up in various ways, and the absence of any arms, does not suggest this. The military hat of this period is shown with three sides symmetrically cocked.

The amalgamation of the Artillery and Engineers had not proved satisfactory. The Artillery resented being in a subordinate position and the Engineers who had served through the War were not yet on an established footing.

A scheme for separating the two arms was put forward by the Chief

FIG. 141. Pioneers, *c.* 1722–24.

Engineer Michael Richards. The Artillery, as has been shown, was made a separate and permanent establishment, and at the same time a regular corps of Engineers was formed consisting of a Chief Engineer, 3 Directors, 6 Engineers in ordinary, 6 Engineers extraordinary, 6 Sub-engineers, and 6 Practitioners. The Warrant for this was given at St. James's, May 26th, 1716.

In 1717 Minorca and Gibraltar also had their Engineer staff, that of the former being 1 Director, 1 Engineer extraordinary and 1 Sub-engineer, and the latter 1 Sub-director and an Engineer in ordinary.

In spite of this separation of the two branches, the Chief Engineer still seems to have been in control, their Warrants placing them in command of the Ordnance trains. Even in 1740 their code of instructions included making lists of numbers, lengths, weights of the different cannon, and how mounted in each garrison, to advise on the number of guns required for a fortress, or field train, the position and building of gun sheds and platforms. Engineers were attached to the expedition to Carthagena in 1741 and a Pioneer detachment was formed of negroes and some white Americans, together with 20 ships' carpenters from the fleet. The negroes appear to have been a failure and it was impossible to get them to work under fire.

Four or five Engineers served with the train in Flanders (1742) and at Fontenoy threw up hasty field trenches and batteries.

A new Warrant was issued Feb. 2nd, 1747, ordering a Brigade of Engineers to be employed. This consisted of 15 Engineers and a Company of miners, with a strength of 2 Lieutenants, 2 Sergeants, 4 Corporals and 34 Miners. The whole of the brigade was present at Lafeldt.

The position of the Engineers was still undefined and was the cause of much dissatisfaction. The Warrants under which they held office being issued by the Ordnance Board, were not recognised as giving them regular military rank. Some attempt to meet this had been made by giving them commissions on the Staff, or in various regiments, but this was not satisfactory as the Colonels of the units to which they were posted would recall them for regimental duty. Eventually, thanks to the efforts of the Duke of Cumberland, the engineers were given military status, their first commissions granted by the King being dated May 14th, 1757.

As in the case of the Artillery, here again is an almost total lack of information regarding the early uniform of the corps. Although now definitely forming part of the army and as such must have had orders as to the uniform to be worn, nothing so far has come to light. Colonel Porter, in his very detailed History of the corps, makes no mention of their early dress. He does, however, give a coloured reproduction of what appears to be a miniature portrait of Gother Mann at the age of sixteen. According to the Army Lists, Mann received no other commission than that in the Royal Engineers, he must therefore be depicted wearing the uniform of that corps.

FIG. 142. Engineer Officer, *c.* 1758.

The late Lt.-Colonel P. H. Kealy, R.E., made endless researches for details of the dress of the R.E., and told me he had found nothing of an early date. In his article on the subject published in the *Royal Engineers Journal* 1934–5, he states: "There are only a few portraits of Engineer officers, of which Mann's miniature is the earliest, and these agree in a general way as to the uniform and that it was a scarlet coat faced with blue with gold lace, in other words the colours of the Artillery dress reversed."

To return to Mann's miniature—the coat is scarlet with a collar of the same edged with gold lace and what appears to be a white patch with a gold button and loop; this, however, has been said to be really the white lining of the coat. The lapels are blue edged with gold lace and having gold loops with buttons at either end, which seems an impossibility, but as the whereabouts of the original appear to be unknown, there are no means of verifying these details. The waistcoat is white, having a gold lace edge and buttonhole loops. The breeches also would be white and a gold laced hat would complete the uniform (Fig. 142).

Since writing the above, thanks to the researches of Mr. Gerald Rickword, we are now able to give the first written contemporary evidence of the uniform of the officers of the corps.

"Army rank granted to corps of Engineers, uniform red lapelled with black velvet buff waistcoats and breeches." (The *Ipswich Journal*, 28 May, 1757.) This agrees with the portrait of Mann, except that this shows blue facings. High lights on black velvet would show a blueish tone; it may also be remarked that very dark blue and black are difficult to distinguish.

It is curious that velvet facings were adopted in the first instance, as they have been retained on the present day full dress. The French army also have black velvet on their full dress.

Armour was still used by Engineer officers during a siege, as a list of stores, Oct. 12th, 1754, mentions: "For the Engineers' armour—backs 3, breasts 3, head pieces 3."

The corps had, as yet, no permanent rank and file, and apparently only small provisional bodies of Pioneers were raised locally. To remedy this need an officer of Engineers drew up a complete scheme for forming companies of Artificers and Pioneers together with drawings of the uniform, arms, equipment and colours to be adopted by the proposed corps. This interesting MS. is in the Record Office, but, unfortunately, the officer does not give his name or the date, merely stating that he had been Chief Engineer for five years in America during the late War. This from the type of dress, etc., of his drawings appears to have been the campaign of 1754–59 in Canada. The Engineer officers who served in America at that time, one of whom may be the author of this plan, were Lt.-Colonel James Montresor, who went out to America with General Braddock as his Chief Engineer in 1754 and served in the same capacity in the attack on Ticonderoga in 1758, Lt.-Colonel J. H. Bushell, who, although senior to Montresor, seems to have served under him at the siege of Louisbourg in 1758; Major Patrick Mackellar went as Engineer in ordinary to America in 1754, served at Fort Du Quesne, at Louisbourg 1758, at Quebec as Chief Engineer 1759–60, and later at Martinique and Havana in 1702 with the same rank. This last officer seems the most probable by the length of his service.

It may be pointed out that the title Chief Engineer was applied to the senior Engineer officer with any force in addition to those holding the actual appointment.

From his proposed scheme, which included ideas for both the settlement and defence of the Colony, the writer appears to have been much in advance of his time, as he recommends the use of breach loading rifles and that the men should be trained in the laying, load and fire. Although his plan does not seem to have come to anything, it eventually materialized in the raising of the military artificers during the siege of Gibraltar 1779–83, which, curiously enough, were formed by the Chief Engineer W. Green, who had served at Louisbourg and Quebec in 1757–59.

In spite of the fact that these companies were never raised, a brief summary of their composition, equipment, etc., are worth recording.

"A scheme for forming the establishment of Engineers into a regular corps, composed of proper artificers, and divided into nine companies,

FIG. 143. Artificer of Proposed Companies.

distinctly trained for the execution of the several works required in camps, sieges and garrison towns."

The writer points out the sum of money that would be saved, as the wages of the same number of hired artificers and labourers, would cost double the amount. That a trained body of military artificers "would be best adapted for service in America, whether employed against the enemy in the woods, or or for securing themselves by proper temporary posts, and for raising or defending, attacking or executing those of greater importance." The companies were to be composed of 2 Captains, 2 Lt.-Captains, 2 Lieutenants, 2 Engineers, 2 Sergeants, 3 Corporals, 1 Drummer and 50 men, an exception being the 1st company, which was to be composed of 40 Cadets, commanded by the same staff of officers and N.C.O.s, and who were to discipline and train the rest of the companies. The Cadets to attend the works of each company for a certain time, and serve at least three years before they can be made an officer, and not then without certificates of being fully qualified for it."

No. 2 Company, Miners; No. 3, Pontoons; No. 4, Sappers; No. 5, Fasciners and Gaboniers; No. 6, Carpenters; No. 7, Smiths; No. 8, Miners; and No. 9, Pioneers.

The staff of the corps: an Adjutant and Quartermaster and Commissary of Stores, and his assistant, a Paymaster, a Surgeon and his mate, Conductors.

Besides these, two Draughtsmen and their assistants, two Surveyors and their assistants, and another Surgeon's mate were to be added to the staff.

The uniform was to be "of a colour not too glaring to be seen at a distance, nor one that is soon soiled, or dirtied, when on works in the field, etc."

The officers, full dress: blue faced and lined buff, waistcoat and breeches of cloth with embroidered gold buttonholes and gilt buttons. Plain hat and gold loop. The officers to have a gorget only when on duty, but no sashes.

FIG. 144. A Officer. B Drummer. C Cadet.

Officers, frock: blue faced and lined buff, buff waistcoat and breeches and plain gilt buttons (Figs. 144 A and 149 A).

Sergeants and Corporals: blue faced and lined buff, buff waistcoat and breeches. The Sergeants to have a narrow gold lace on waistcoats, buttonholes only. The Corporals plain, but to have a small bow knot on the shoulder (Fig. 145).

Cadets as above, but the waistcoat to have the buttonholes embroidered only (Fig. 144 C).

Drummers: blue faced buff as above, but the seams to be laced with a blue and buff coloured lace and the breeches of buff leather (Fig. 144 B).

FIG. 145. Corporal.

Artificers and Pioneers: blue faced buff, buff leather waistcoats and breeches, proper to perform their work in. To have two pockets on each side, the uppermost to be elbow height, that on left for cartridges, the other for spare flints, etc., for use of their arms. The two undermost for small tools or implements, belonging to the profession of the Company, to be always ready and at hand when wanted for repairs or on a march, etc.

The whole to wear a leather cap lined with fur, to be worn next the head in winter, with its cape to button under the chin, and fall over part of the shoulder, to prevent cold, rain and wet. In summer the leather to be next the head, the fur being made to take out on occasions. The cap was to have a yellow tuft or feather. The men were to have hats besides, for full dress apparently, as shown in sketches. They were to have buff canvas stockings for review and black leather leggings on a march, duty or work (Figs. 147 and 146).

FIG. 146. Artificers.

Arms: Officers to carry a spontoon or small partisan of six feet, including point of spear and ferril. The spear to be ten inches long and two and a half inches broad at the base. The above to serve as a measure of 6 foot or toise, and also as a defensive weapon.

The Sergeants, Corporals, Cadets, Artificers and Pioneers to have muskets of no more than 2 ft. 8 ins. length in the barrel, which is to be rifled and to be loaded at the guard, which will avoid the troublesome, dangerous and noisy motion of drawing and returning their rammers before an enemy, besides the advantage of loading on their backs and firing on their bellies without being exposed (Fig. 148).

FIG. 147. Artificer's Cap.

Each Artificer and Pioneer to have the tool mostly used in his Company, that is convenient to mount on a socket, for fixing on occasion to his musket, and carried sheathed as necessary and side arms.

That all commands in the field, march, or in danger of meeting the enemy, to carry their muskets shoulder'd but otherwise slung.

The manual exercise to be taught in shortest and most concise manner.

FIG. 148. Artificer with Breech-loading Musket.

"Practise of firing dextrously at a mark standing, kneeling, and almost lyeing on the ground. To be taught to load quick on their backs and at one time. To fire together by signals in the day and by particular whistles in the night, as their services may require it often" (Figs. 143 and 148).

"Colours proposed for this Corps: That is case of a review or the whole corps or major part being together, it is necessary there should be two colours mounted on a staff 8 ft. long. The fly to be two feet square each, on one side the union and crown, and on the other the ordnance arms with a Phœnix underneath with the motto *Ex Flaminis Resurexi*. The above colour always to be at the head of the encampment."

The two sketches show a red ensign with the union with white scrolls either side and the crown above. The other ensign is blue with a shield having the ordnance arms, three yellow guns on a blue field and three blue cannon balls on white or argent. The shield is edged with yellow and below a yellow ribbon with the motto and the sphinx below (Fig. 149 B).

FIG. 149. A Officer. B Colours and Gorget of Artificer Corps. D Private.

GENERAL OFFICERS

1715–1760

GENERAL officers in the time of George I are often depicted in their portraits wearing plain scarlet coats with round cuffs of the same colour. Sometimes the buttonholes have loops of gold lace, in other cases the coat is embroidered down the front edges and pocket flaps with gold. These portraits show the sitters wearing the full curled and powdered wig, but smaller than those of the previous reign.

FIG. 150. General Officer, Early 18th Century.

FIG. 151. General Officer, Early 18th Century.

The portraits of the two Generals (Figs. 150 and 151) are good examples of the different types of the dress of General officers of the beginning of the 18th century; there is nothing to indicate their rank. The change in the style of dressing the wig, which was taking place, is well shown.

Later when lapels became the fashion the uniform in which Generals are shown is a scarlet coat with blue lapels and round cuffs without a slash. The lapels have large gold lace loops to the buttonholes and the blue cuffs have one or more rings or edgings of gold lace, the whole closely resembling the full dress of the Foot Guards; in fact it has been suggested that it was this uniform which was worn by General officers (Fig. 153).

FIG. 152. General Blakeney, 1755.

The engraving of the portrait of General Blakeney, painted at Minorca in 1755, shows him wearing this type of uniform. Blakeney, at the age of eighty-four, was the gallant defender of Minorca. For his failure to relieve the garrison of this fortress, the unfortunate Admiral Byng was executed, not as Voltaire remarked "*pour encourager les autres*", but to protect the Ministers from the popular outcry over the loss of the place; to further which ends "the enquiry" into the loss of Minorca was snowed under by "cart loads of papers and such bushels of letters that the very dates fill three and twenty sheets of paper" (Fig. 152).

The waistcoat and breeches are buff. The sash is mostly shown worn round the waist and said by some authorities to have been the mode adopted by General officers.

There are two portraits of the Duke of Cumberland by Morier which give two good examples of this dress and which corresponds fairly closely with the portraits of many other Generals. In the first the Duke wears a scarlet coat with blue lapels with wide gold lace buttonholes, buff waistcoat,

FIG. 153. General Officer, *c.* 1750.

also edged with gold lace, and buff breeches. On the right shoulder is a gold shoulder knot or aiguillet and a gold fringed crimson sash round the waist. He carries a crimson gold-mounted baton. The housings are blue velvet embroidered with silver and are deeply fringed along the edges with the same. In the second picture the Duke wears what seems to be the frock or undress uniform. The lapels blue edged with gold but without the wide lace buttonholes, the waistcoat and breeches are buff; the sword knot is gold and crimson and the sash is plain crimson (Figs. 154 and 155).

The housings are scarlet with wide gold lace edging; on the front and rear corners of the housings are the loops and heavy gold tassels with two smaller tassels, above these knots are crescents. These crescents are frequently shown on the housings of Colonels of regiments and one can only suppose that they originated like the crescents on the head stalls, etc., from the horse trapping ornaments of the East.

FIG. 154. The Duke of Cumberland and Orderly, *c.* 1750.

FIG. 155. The Duke of Cumberland.

In both portraits the Duke wears a cuirass. This is quite usually shown in portraits of Generals, but there has been some doubt whether it were worn in the field by officers. There is one piece of evidence, however, in support of its being used. An officer of the Gardes Françaises at Fontenoy mentions that when his men were thrown into confusion by the deadly platoon fire of the Foot Guards he was thrown down by his men and was only saved by his curasse from being trampled to death.

These paintings and those of other officers show the new fashion of dressing the hair, which is brushed back from the forehead and twisted into a long queue, or pigtail, which was lengthened by additional material until it reached almost to the waist. The regulation in Germany and France at one time was that it should terminate between the two waist buttons on the back of the coat. One or more side curls were worn on each side of the head. The hair was powdered, but not in every case, and was tied at the back with a bow of black ribbon, part of which was also used to prevent the coat from being stained by the hair grease. This ribbon, as everyone knows, is still perpetuated by the black ribbon flash worn by the Royal Welch Fusiliers at the back of their tunic collar.

Although no orders, as far as is known, had yet been issued concerning General officers' dress, there are some portraits which show the laced button holes arranged in groups of twos, threes or fours instead of at equal distances, and one wonders if this was not already a commencement of the way to distinguish their rank which was adopted later, otherwise the same style of uniform is shown in most of the paintings so far as full dress is concerned, although there is evidence that a plain red or blue frock was worn. The well-known portrait of Wolfe at Quebec by J. S. C. Schaak, the details of which were based on a sketch made at Quebec by Captain Harvey Smythe, one of Wolfe's A.D.C.s, shows the Major-

FIG. 156. General Wolfe at Quebec.

FIG. 157. General Wolfe's Campaigning Cloak.

General in a plain red coat with red cuffs, red waistcoat and breeches, the latter being coloured a dark blue tone in the original sketch. The boots are supple black leather with brown tops or turn-overs.

The hat like the rest of the uniform is without any lace and its only ornament the black cockade. The hair is unpowdered, showing the natural red colour, and is tied in a long queue. A black band is tied round the left arm. The equipment consists of a waist belt with a frog holding the bayonet, but no sword. The fuzil is slung over his shoulder. The whole represents a practical service kit and was no doubt similar to that worn by the officers in the forest campaigns of this period.

FIG. 158. A Cumberland Hussar.

The Duke of Cumberland had two Hanoverians as orderlies. These were dressed in a Hussar uniform of his livery, crimson and green, consisting of a green jacket bound and laced with white, crimson collar and cuffs, crimson breeches, a tall Hussar cap of brown fur with a crimson bag, black Hessian boots with a tassel, a crimson sabretache edged white and embroidered with a ducal coronet and letter W within the Garter. The housings are crimson edged white and having the same device as the sabretache embroidered on the rear corners. These Hussars are shown by Paul Sandby in his various sketches of the Duke. Mostly they are too small to show the details very clearly—but the larger figure (Fig. 158) shows the details more clearly. In this sketch, however, the boots are similar to boots worn by dragoons. The silver lace and crimson cuffs are more distinctly shown. In one of Sandby's sketches one of these Hussars is represented wearing a crimson waistcoat.

The Germans were recognized good horse masters and no doubt for this reason the Marquess of Granby had a Hessian orderly, John Notzel, who is shown holding the General's charger in one of his portraits by Sir Joshua Reynolds, who however calls him a Swiss. He, also, is wearing a Hussar uniform of a white dolman and blue pelisse, fur cap and red Hessian boots.

FIG. 159. The Marquess of Granby and his orderly.

1714–1760

THE Association of the Honourable Artillery Company with the Train Bands was still continued by its supplying the latter with officers.

A detachment of the Company marched in the procession when George I made his state entry into the City and the following order for the dress to be worn on this occasion was issued: buff coats; gold laced hats with red feathers; "the hair to be tied behind with red ribbonds and with black ribbonds before; white stockings, black garters, fuzees unslung, and cartouch boxes." The red feather which had for so long been the distinguishing feature of the regiment disappeared about 1723.

FIG. 160. Officer's Grenadier Cap, H.A.C., c. 1716.

The Grenadiers wore a cap of crimson velvet very similar to that worn in the preceding reign, but the front was now embellished with the full arms and supporters on a stamped and cut out metal plate, the only instance that is known in England of a cloth Grenadier cap being ornamented in metal. The cypher on the little flap is that of George I (Fig. 160).

The officers of the Train Bands were admitted to the Company without fees, but were to continue to act as Sergeants and wear buff coats.

Half pikes only were to be carried on the general marches held by the Corps.

King George I reviewed the Company in St. James's Park in 1722, and from the orders for this occasion we learn that scarlet was worn for the first time since the Civil War. "All officers and those armed with half pikes to wear scarlet coats, white stockings, and black garters, the Fusiliers to wear buff coats, laced hats, hair in bags, white stockings or spatterdashes and black garters."

"The commissioned officers to wear white feathers in their hats, and the flankers red, or none."

"All mounted officers to dismount on the parade ground and to march past on foot," which was the usual procedure at the time.

The review seems to have proved a great success and the appearance of the scarlet coats so satisfactory that it was ordered that, when reviewed by the King, "all officers and half pike men in scarlet and Fusiliers in buff. The Grenadiers, both officers and men, to continue to wear buff."

Another review of the Company was held on 2nd March, 1727. The orders for dress were the same except that commissioned officers were to have laced hats with black cockades instead of white feathers. The captain of the Pioneers to carry a field staff and his Lieutenant a battle-axe. These were ordered to salute by doffing their hats only.

The whole order for the march past was laid down in full detail:

1st. The Captain and his Pioneers with their accoutrements.

2nd. The servants, grooms and horses of the Grenadier officers.

3rd. The Marshal of Grenadiers.

4th. The Grenadiers' music.

5th. The Captain, Lieutenants and Grenadiers.

6th. The Led horses of the general officers with grooms on horseback.

7th. The servants of the general officers.

8th. The Marshals of the general officers.

9th. The Beadle of the Company.

10th. The Music.

11th. The General, the Lt.-General, three Captains with the Captain-Lieutenant.

12th. Fusiliers in Buff with Flankers in scarlet and sashes round their waist and partizans shouldered.

13th. The Surgeon with his scarf and arms of the Company.

14th. Three Ensigns of the general officers.

15th. The Pikes in scarlet with Flankers as before.

16th. Three Captains-Ensigns.

17th. The second body of Pikes.

18th. The second body of Fuziliers in buff with various officers bringing up the rear.

Three Sheets of Colours. Artillery First Division. Captains. Grenadiers.

FIG. 161. Honourable Artille

FIG. 162. Officer H.A.C., *c.* 1726.

An engraving of the Lord Mayor's procession, *c.* 1742, shows the various branches of the Honourable Artillery Company (Fig. 161).

The figures are of very small dimensions so that it is impossible to distinguish minor details and colouring, but, even so, we are fortunate in having pictorial evidence of the band of music, pioneers, matrosses, etc. of this historic corps.

The presence of the men with the "Holy water sprinklers" is particularly interesting. These were carried by the corps from an early date, though it is not clear what was the use of these mediæval weapons. It may be noted that they are shown in the trophies of arms painted on the colours of the Ordnance Regiment (The Royal Fusiliers), 1685, and in other early decorative designs symbolic of artillery, but whether this indicates that they were part of the equipment of the train of artillery, or merely part

wo Lieutenants. Captain. Marshal. Music. Lieutenant. Pioneers. Captain.
mpany. 1742, Oct. 25

of the design of warlike emblems, it is impossible to say.

The head-dress of the Pioneers appears to be the cloth cap with loose hood or bag.

Although the pike had long since been abolished the Company seem to have retained the half pike for ceremonial purposes.

A series of drawings by William Hogarth showing the salute with half pike and halbard, depict an officer and Sergeant of the Company in 1726. These drawings were apparently done for a drill book written by Captain John Blackwell of the Company. These plates were later reproduced in Grose's *Military Antiquities* (Figs. 162 and 163).

From the various dress orders mentioned above we may presume that the officer wears a red coat and the Sergeant a buff one, the latter is also laced down the seams, the usual distinction of his rank.

FIG. 163. Sergeant H.A.C., 1726.

VOLUNTEER CORPS

1745

Besides those regiments which were additional units of the regular army, many volunteer corps were formed during the Jacobite rising of 1745. From the fragmentary references to them it is at times difficult to decide if they apply to the same corps or to different units. By their constitution they may be considered as the precursors of the volunteers of the seventeen-nineties and eighteen-sixties. Under the incentive of the Archbishop an association for the defence of the city was formed at the Castle of York, September 23rd, 1745. It consisted of 800 of the principal nobility and gentry of the County.

John Roper, the Lord Mayor, also convened a meeting at which subscriptions were raised for the forming of four companies of seventy men each exclusive of Sergeants, Corporals and Drummers, entitled the Yorkshire Blues, the officers serving without pay. These companies remained embodied for four months.

Fig. 164. Captain Thornton.

Captain Thornton, later Colonel of the West Yorkshire Militia, raised a company at his own expense and was present at Falkirk. This company appears to have formed part of the Yorkshire Blues and it is said he is wearing the uniform of the corps in his portrait. If this be so, it consists of a blue coat lined blue, red turn-down collar, lapels and cuffs with narrow gold buttonholes and gilt buttons, buff waistcoat and breeches. The boots come to below the knee and are somewhat Hessian in shape with a red edging round the top (Fig. 164).

Yet another body called "The Independents" was formed for the defence of the city of York by the gentlemen and principal inhabitants, their uniform and accoutrements being all bought at their own expense. The corps remained on duty for ten months.

A body of Foxhunters under General Oglethorpe was formed into a regiment of Light Horse, or what would now be called a Yeomanry Corps.

A portrait of Colonel Braddyll by Romney is described as wearing a Yorkshire Yeomanry uniform and may, it is said, be that of this corps; long red coat, gold lace on the collar and cuffs, gold cord loops for buttonholes, three-cornered cocked hat and boots nearly to the knees with brown turn-over tops. This picture however was painted many years later.

The Volunteer regiments of Exon and Devon were given blue cloth coats faced up with red and lined with red long ell. The hat was edged with white worsted lace and a white metal button. The like buttons on the coat and a pair of white garters to each man. (*State Domestic Papers*, Geo. II, Bundle 72, No. 134.)

"The Independent Companies of Gentlemen Volunteers raised for our service within our city and suburbs of London" seem to have been quite numerous and some to have been the strength of a regiment as "The Regiment of Gentlemen Volunteers of London" under Colonel Michal Delaboles, Lt.-Colonel Joseph Otway and Major Theophilus Bancos, The Tower Hamlets Volunteers raised by Lord Cornwallis, The Independent Company of Royal Volunteers of Captain Cosman.

A Company of Gentlemen Volunteers known as the Loyal Blue Fusiliers from its uniform, after the scare was over, joined the Honourable Artillery Company under the condition of remaining a separate division.

Another body was The Volunteers of Leathersellers Hall, so called from their drilling in the Hall of that Guild. From an allusion to the dress of their Drummers it appears they wore red coats with wings.

A print in the Royal Library at Windsor entitled "The Loyal Associators" shows this corps marching through the streets of London. They all wear single-breasted coats without any lapels or lacing, large plain round cuffs without any lace or slash, laced three-cornered hats and black cockades. The officer wears a gorget and a sash over the left shoulder and carried a spontoon. At the top of the picture is the device of clasped hands within a circle formed by two serpents.

For the training of these corps a small book was produced entitled *The Gentleman Volunteer's Pocket Companion*. It contains a series of figures performing the musket manual. These have often been considered on account of the title of the book to represent the Volunteers, whereas they are, in reality, men of the 1st Regiment of Foot Guards.

ENGLISH MILITIA

1758–1760

THE Militia seems for long past to have gradually fallen into a moribund state; musters were no longer held and no training appears to have been enforced, in fact the force appears to have existed in name only. The mounted Militia seems to have disappeared at the beginning of the eighteenth century, as in 1715 "armour for horse" was ordered to be abolished. As horse armour had ceased to be used for nearly a century it seems probable that this is a clerical error and what is meant is the disbanding of the Troops of Militia cavalry.*

FIG. 165. Militia Officer, Early 18th Century.

With the Jacobite scare some units seem to have been hastily embodied for a short time, but such was the lack of local troops in the country that Hanoverian and Hessian regiments were brought over for garrison duty. The following appears to refer to the East and West Riding Militia:

A letter, dated 1745, to Grimston at Beverley, Wakefield, *re* the clothing of his Company, states that "blue kersey is to be used for the coats and red kersey for the facings, bays for the linings. Two dozen buttons for each coat. The tailors to find thread and tape. The West Riding coats are double-breasted."

FIG. 166. Officer, Lancashire Militia.

* Or this may apply to the cuirasse of the men.

being edged and looped with lace.

Kent. Scarlet coats, sky blue lapels, cuffs and waistcoats, both laced and looped with silver.

Leicester. Scarlet coats, white collar, lapels and cuffs with a red slash, the buttons are gilt and the buttonholes stitched with gold thread. The portrait shows a turn-down collar, which is different to the customary narrow scarlet collar of this period.

FIG. 171. Officer, Leicester Militia.

Monmouth and Brecon. Green facings for both and the portrait of Thomas Morgan shows lapels, cuffs (and slash) and waistcoat all laced.

The Norfolk Militia are described in the *London Post* and *London Chronicle* of 1759 as wearing red faced with black. "The Drummers and fifers were all little boys with fur caps and looked very pretty."

FIG. 172. Norfolk Militia, 1759. "Rear rank make ready."

A drill book entitled *The Norfolk Militia* published in 1759 has pictures of officers and men performing the manual, etc. Both appear to have coats without lapels, those of the officers having the buttonholes laced with gold. The officers' coat facings were of black velvet.

The Lincoln Militia are said to have been granted the title of Royal for service in Liverpool when Thurot alarmed the North-west coast in 1760 and always used the title although not officially confirmed until 1809.

FIG. 173. Officer's coat and waistcoat, East Yorks Militia, *c.* 1760.

The 1st Devon wore yellow facings and the North Devon green.

The three Ridings of Yorkshire, East, North and West were treated as separate corps as far as Militia organisation was concerned. An officer's coat of the East York is preserved in the Leeds Art Gallery. It is scarlet with buff lapels, cuffs and waistcoat laced with silver.

The chapter on clothing in the *History of the King's Own Regiment* gives the following details of the uniform of the Lancashire Militia, 1758, from actual specimens in the possession of the regiment:

"*Officers*. Hat laced gold, coat scarlet with low red collar with blue patches and brass buttons. The pockets are cut obliquely and have four brass buttons; lapels, skirt slits and front of skirt are edged with gold lace. The lapels, cuffs and linings are blue. The lapels have ten gold laced loops and buttons in pairs on each lapel. The cuff slash has four loops and buttons. The buttons are dome-shaped. The breeches and waistcoat are scarlet and the latter has eighteen gold lace loops and buttons; white gaiters with black bone buttons and black garters."

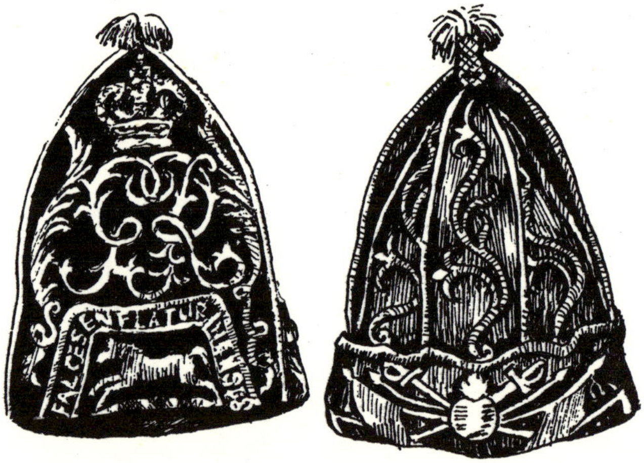

Fig. 174. Officer's Cap, Hampshire Militia.

"*The Rank and file*. Plain hat, red coat faced, lined and lapelled blue; red waistcoat, blue breeches, white hose and black garters."

The Militia, like the regular regiments, had their Grenadier Companies and one or two caps are in existence. That of the Hampshire Militia is 12½ ins. high and of crimson velvet with a stiff silver tassel. The front and little flap are of black velvet, the latter having the motto on a black label, *Falces conflantur in Enses*. The front has the cypher in silver with side scrolls of gold, the whole surmounted by the crown proper. On the turn-up at the back is a large grenade in the centre of a trophy of red flags, swords, axes, etc., worked in silver as are the scrolls and seams on the crown.

Another interesting Militia cap is that of the Cumberland Militia in the collection of Mr. Catley, by whose permission I am allowed to make this illustration and to quote the description from his article in the *Journal for Army Historical Research*.

FIG. 175. Officer's Cap, Cumberland Militia.

The cap is red cloth; the front and turn-up at the back are blue; the tuft is silver and crimson. The front, edged with silver, is embroidered with the Lowther arms, crest and cypher; a silver dragon gorged and chained silver. The orle is silver and black. Cypher in silver. In the centre is a shield bearing a red hand on a white field. The shield is edged with gold thread-work and the floral scroll design on either side of the cypher is worked in silver.

The little flap is red embroidered with a gold grenade and scroll and leaves in silver.

The back is embroidered with a floral scroll edged with yellow thread. The grenade is gold. All the embroidery on the back is silver stitched with yellow thread on the edges, except the grenade which is gold (Fig. 175).

At Redworth Hall is a most interesting fur cap of the Durham Militia as well as the portrait of Crozier Surtee in the uniform of the Corps; a similar fur cap is shown in the painting (Figs. 176 and 177).

The following details have been supplied through the courtesy of Major Henry S. Surtees, the present owner:

A scarlet coat with *blue* cuffs and lapels, silver lace gorget and epaulet, buff waistcoat and breeches.

The fur cap is closely embroidered on the front giving it the effect of a "plate"; at the apex is a fist holding a broken sword issuing from a coronet. In the centre are the letters D. M. (Durham Militia) in raised embroidery,

Note the discrepancy between colour of facings and cap is again a case of a change of tone. It seems that the actual facings were green.

with the motto *Nec Temere nec Timere* round the edge. On either side are a griffin and a unicorn as supporters. The turn-up at the back is green and the hood or bag is a yellowish green tone with a network of silver and a tassel of the same. It is probable that the velvet was originally a more definite green colour. The button-holes on the cuffs and skirts are more of the cavalry pattern than that of infantry, but there are other cases of Militia regiments having adopted this style.

FIG. 177. Officer's Cap, Durham Militia, *c.* 1760.

FIG. 176. Grenadier Officer, Durham Militia, *c.* 1760.

It is curious that only one epaulet is worn instead of two, as was usual for an officer of the Grenadier company.

Crozier Surtees received his commission in 1759 at the age of twenty and was a man of very large stature.

FIG. 178. Norfolk Militia, 1758. "Present, Second and First Rank."

MILITIA COLOURS

FIG. 179. Ensign, Norfolk Militia, 1759.

The regimental, or 2nd, colour seems to have been the facing colour and in many cases had the arms of the county or those of the Lord-Lieutenant. The list appended compiled from Ordnance Warrants W.O. 55; 427 and H.O. 51; 443 mentions a certain number of the colours from which the facings can also be learned. Unfortunately there are many blanks as to the colour of the 2nd colour and also as to who was the Lord-Lieutenant.

County.	*Date.*	*Regimental Colour.*
Bedford	1760	Dark green sheet with Arms of
Berks	1759	Blue Sheet with Arms of Duke of St. Albans.
Bucks	1759	Arms of County of Bucks in centre and Arms of Lord-Lieutenant in one corner.
Cambridge	1759	Green sheet with Arms of Viscount Royston.
Cheshire	1759	Arms of Earl of Chester.
Cornwall	1760	Black sheet with 15 gold balls, being arms of Cornwall; Lord-Lieutenant's Arms, small in corner.
Cumberland	1760	Red sheet with crest and cypher of Sir James Lowther in blue.
Durham	1760	Arms of Lord-Lieutenant on a sheet.
Gloucester, S.	1759	Arms of on blue sheet.
Gloucester, N.	1761	South Battalion device on blue sheet.
Hants	1759	Arms of Lord-Lieutenant.
Herts	1759	Arms of County on buff field.
Huntingdon	1759	Black sheet with Arms of Duke of Manchester.
Lancashire	1760	Arms of Duke of Lancaster on blue sheet.
Leicester	1760	Arms of Duke of Rutland.

County.	*Date.*	*Regimental Colour.*
Lincoln, N.	1759	Red sheet, with Arms of Duke of Ancaster.
Lincoln, S.	1759	Red sheet, with Arms of Duke of Ancaster.
Middlesex, E.	1760	Duke of Newcastle's Arms on white sheet.
Northampton	1763	Arms of Lord-Lieutenant on a sheet.
Northumberland	1759	. . . sheet with Arms of Earl of Northumberland.
Rutland	1759	Pale buff sheet with Arms of Earl of Exeter.
Shropshire	1762	Arms of Lord-Lieutenant.
Somerset	1769	Lord-Lieutenant's Arms.
Suffolk	1759	Lord-Lieutenant's Arms.
Surrey	1759	Lord-Lieutenant's Arms.
Westmorland	1759	Red sheet with crest and cypher of Sir James Lowther in blue.
Yorks, E. Riding	1759	. . . sheet, Arms of Lord-Lieutenant.
Yorks, N. Riding (Richmondshire)	1759	Blue sheet with Arms of Earl of Holdernesse
Yorks, W. Riding	1759	Green sheet with Arms of Lord-Lieutenant (Marquis of Rockingham).
Yorks, W. Riding (2nd Battalion)	1759	Green sheet with Arms of Lord-Lieutenant (Marquis of Rockingham).
Yorks, N. Riding (Cleveland and Bulmer)	1759	Blue sheet with Arms of Earl of Holdernesse.
Anglesey	1762	Blue with Arms of Lord-Lieutenant.
Brecknock	1759	Green sheet with Arms of . . .
Carnarvon	1762	White sheet with Arms of Lord-Lieutenant.
Carmarthen	1759	Blue sheet with Lord-Lieutenant's Arms.
Denbigh	1760	Arms of Lord-Lieutenant's on a . . . sheet.
Flint	1759	Lord-Lieutenant's Arms on green sheet.
Glamorgan	1760	Red sheet with Arms of Lord-Lieutenant.
Merioneth	1762	Arms of Lord-Lieutenant's on a . . . sheet.
Monmouth	1759	Green sheet with Arms of Lord-Lieutenant.
Pembroke	1759	Both colours blue, Arms of County on one, Arms of County Town on other.
Montgomery	1763	Arms of Earl of Powis.
Radnor	1763	Blue sheet with Arms of Lord-Lieutenant.

JERSEY MILITIA

1714–1760

The 1744 regulations for the Militia consist of 26 articles and include the following:

Anyone sighting the approach of an enemy fleet or who heard two cannon fired in the island will at once raise the alarm, muskets will be fired, beacons lit, and bells rung. On the alarm being raised everyone was to hurry to his allotted place of assembly.

One Troop of Cavalry was to meet at St. Peter's Church, the other at Grouville Windmill.

The infantry and Parish gunners to fall in at their Parish churches. The artillery to man their respective coastal batteries and await orders from their group commanders.

There were three groups, the East Coast, the shore batteries at St. Aubin's Bay and those at St. Ouen's Bay.

There were 14 Guard houses along the coast, their garrisons amounting to 10 officers and 185 men.

At Guards commanded by officers the drums were to beat the "Taptow" one hour after sunset and the réveillée at dawn. The whole coast line was kept under observation day and night by guards and patrols.

"24 Sept. 1744, Lord Cobham Governor.

"New Regulations for the Militia. To the end that the Militia may appear uniformly dressed: it is ordered that all officers and soldiers shall appear on all parades in red coats and white stockings and that each soldier shall be provided with a good musket, sword and pouch with the requisite ammunition all in good order on pain of being judged a defaulter."

IRISH MILITIA AND VOLUNTEERS

THE Irish Militia appears to have started on much the same lines as in England and Scotland, independent Corps of one or more companies of foot being raised, sometimes with a Troop of Horse in addition. In towns like Dublin, Cork, etc., these forces seem to have been of the strength of a regiment. Unfortunately very few records of the force now exist, as these were no doubt destroyed in the burning of the Four Courts in 1922. It seems unlikely that much information will now be discovered. As in Scotland some of the units may have worn coats of homespun.*

The Inniskilling Regiments of 1689 from rather vague references seem at first to have worn some sort of grey kit before receiving their red coats sent out from England. No doubt some of those regiments raised in the cities wore the red of the regular army; later, blue seems to have been adopted by certain corps.

The Council Book of the Corporation of the City of Cork gives the following fragmentary notices of their Militia, which as far back as 1650 was in existence under the title of the Train Band of Cork.

1690, Dec. 12th. "That whereas the Militia of the County is to be regimented and as the Militia of the town is at present but 4 Foot companies and 1 Troop of Horse, too few for an entire regiment, and to prevent our being regimented under a county Colonel, that this city and county thereof raise 4 Foot companies which will amount to 8 companies and 1 Troop, each company to be of 60 men and the Troop 50. The Mayor of Cork for the time being to be Colonel."

1691, Jan. 28th. "A muster master of the Militia Horse and Foot appointed."

1708. "350 firelocks received 1709. Muskets to be repaired at the charge of the city and 6 chests of arms sent to the Militia."

1715, Oct. 27th. "That the Corporation provide 3 pair of colours for the militia." Oct. 30th. "230 Dublin muskets and 120 English muskets for the militia to be drawn from H.M. Stores, Dublin."

1716, Feb. 16th. "That a Corporal and 8 Horsemen of Militia patrol

* Among the items of clothing for the Irish Militia, sent from England in the Late 17th Century, crimson sashes are mentioned.

North and South suburbs at 6 P.M. at night. The Mayor to provide horses."

1720. "Payments for making bayonets and for supply of ball and flints."

1745, Oct. 17th. "That the commissioners of array are of opinion that, to execute the orders of the Government for training and exercising the Militia, it is necessary that a sum of money be raised to pay for colours, drums, standards, halberts and grenadier caps and that the aforesaid be provided."

1762, Aug, 21st. "Muskets, bayonets and hangers to be received from H.M. stores."

1763, Jan. 7th. "As peace with France and Spain is agreed upon the Corporation are ready to return arms to Stores."

Probably owing to the Jacobite rising a Corps entitled "The True Blue" was raised in Cork. The cavalry were dressed in blue coats laced with silver and having silver shoulder knots and buttons. The horse furniture consisted of goat skins. The infantry consisted of 4 companies, 1 of grenadiers, 1 Light company and 2 Battalion companies. The presence of a Light company is interesting as there is no mention elsewhere of such being formed at this time. The Corps was commanded by Colonel Richard Lord Shannon.

On May 13th, 1659, the inhabitants of Dublin were ordered to be formed into 2 regiments, one within the city of 9 companies, and one in the suburbs of 10 companies, the Mayor, Recorder, Sheriffs and Aldermen to be commissioners for the Militia.

The lists of Dublin Militia in Faulkner's *Dublin Journal*, 1744–1764, give 1 regiment of Horse and 3 regiments of Foot.

The City of Waterford in 1746 had a regiment of Foot 500 strong in 10 companies, including one of grenadiers. The uniform was a blue coat faced scarlet, gilt buttons, scarlet waistcoat and breeches, and a gold laced hat. There was in addition an independent Troop of Horse dressed in much the same uniform.

The Grenadier company and the Troop of Horse were composed of men of some position and willing to pay for their regimentals and appear to have been the most uniformly clothed part of the force.

Belfast raised a corps of Volunteers in 1757 called "The Belfast Young Volunteer Company." In 1760 owing to the threat of a French invasion various Volunteer units besides Militia were raised in the counties Antrim, Down and Armagh, and a little later, owing to the "Whiteboy" outbreaks further Corps were raised for the preservation of order in the country.

THE initiative for assistance being given to old soldiers who had suffered in the service of the country was due to Queen Mary I, who in her Will dated 1558 left instructions that a house in London should be endowed with a yearly sum of 400 marks "for the relief and help of poor, impotent and aged soldiers."

In 1680 the Royal Hospital, Kilmainham, Dublin, was founded and the Royal Hospital, Chelsea, the following year.

The Chelsea Establishment has always been associated with Nell Gwyn,* one story being that it was due to her using her influence with Charles II in support of the proposed foundation; another, that it was actually her suggestion that a home should be provided for old and maimed soldiers, and yet another version being that the land selected for the site had already been granted to her by the King and that she willingly waived her claim for so good a cause. Whatever the actual facts, by many of the Old Pensioners Nell Gwyn was regarded almost in the light of the founder, one veteran pointing out one lady in the allegorical painting over the altar as being her likeness.

The uniform given to the Chelsea Pensioners was in its general outline except for minor details and difference in cut much as it has remained to this day.

In 1692 the Pensioners were divided into four grades, viz., officers, cavalry, sergeants and privates.

The dress of the corporals, drummers and privates in that year was:

"A red cloth coat with brass buttons, lined with blew bays and false sleeves with a large cypher of the King and Queen on the back. Blew kersey breeches lined with brown orsenbriggs, £1 14s."

"Two black hats with copper edgings and white hair hat bands at 5s. 6d. each = 11s.

"Two pairs of stockings, 3s. 2d.; two pairs of shoes, 8s.; a pair shoo buckles, 3d.; four shirts, 10s.; 4 neckcloths, 2s. 8d.; two night caps, 10d. Total, £3 10s. 7d."

"For the cavalry and sergeants a crimson cloth coat with large brass buttons lined with blew serge, the sleeves turned up with ye same cloth and bound with gold braid, a pair of breeches of blew broad cloth lined with white fustian, a black hat with gold edging and a white hair hat band."

* Captain C. G. T. Dean in his excellent *History of the Hospital* says: "There is no mention of Nell Gwynne's name in connection with the Hospital until 1752 when an entirely fictitious pamphlet of her life was published."

Later the sergeants had silver lace and pewter buttons instead of the gold and brass items of the cavalry.

"The officers' uniform was a scarlet cloth coat with gilt brass buttons lined with blew shalloon, the sleeves turned up with the same cloth and bound about with gold braid and a pair of breeches of fine blew cloth lined with white fustian, a black hat edged with the same lace as ye sleeves and a gold hat band."

Surtout coats were provided by the terms of a legacy from the Earl of Ranelagh from 1712 onwards. These were at first of light blue with facings of a darker blue. This seems to be the origin of the "Hospital blue."

About 1718 the uniform of the corporals and privates was "a red cloth coat lined with blue bays, with brass buttons, a pair of kersey breeches, a waistcoat of blew kersey with brass buttons, a black hat with copper edging and white hat band," and a little later a black cockade was added to the hat.

A portrait in the Great Hall of the Hospital shows a pensioner of 1730 in this uniform—the red full-skirted coat and blue cuffs and blue waistcoat, very similar to the figure of the Invalid in the 1742 Clothing Book and Morier's painting.

The officers' dress in 1718 was "a scarlet cloth coat lined blew shalloon, the sleeves turned up with the said cloth with gold thread buttons and button-holes, a pair of blew cloth breeches, with woolen Rolls at the knee or buttoned, waistcoat of blew cloth skirted with blue shalloon and blew thread buttons, a black hat edged with gold and black twist hat band."

The Pensioners were originally armed with musket, bayonet and sword. These were stored in a small cabin-like room at the end of each ward. The rooms are still to be seen, but are now used as sleeping compartments for the sergeants. It should be remembered that those Pensioners capable of bearing arms were on occasion called upon to serve again, generally on garrison duty, some being even drafted as Marines into ships of Anson's squadron.

Of the uniform of the Kilmainham Establishment nothing much is known except that the Pensioners wore a similar dress of red faced with blue.

The following lists give a little supplementary information:

1702, June 24th. For the men "a coate & waste coate, 1 hat, 1 payre of stockens, 2 shirts, 1 cravatt, 1 payre of shoes & 2 linnen capps."

1710, December 11th. "Each officer to be furnished with a scarlet coat, and Athol hat laced with gold lace & a payre of blue worsted stockings."

THE salute was originally made by both officers and men by doffing their hats. In a French Drill Book of 1697 in the writer's collection, the officers of the Gardes Françaises are represented as bowing at the same time, which, no doubt, would be the universal custom and quite in keeping with that courtly period.

FIG. 180. Officer Saluting.

A book entitled *The Soldier's Pocket Companion* was published in 1745, but the illustrations, as is often the case, appear to be of an earlier date, as the officers are depicted with the half pike instead of the spontoon which replaced it. The illustrations show officers in the various postures of the salute, the half pike being held in the right hand horizontally on the shoulder while the left carries the hat. (Figs. 180 and 181).

Hogarth drew a series of figures of officers and sergeants of the Honourable Artillery Company performing the salute with half pike and halbert for a Drill Book entitled *A Compendium of Military Discipline as it is Practised by the Hon. Artillery Company*, etc., etc., published in 1726. The postures are very similar to those in the first-mentioned book (Figs. 162 and 163).

The Order Books of the 1st Foot Guards have two entries for June, 1750, and May, 1753, which state that "all soldiers are to take off their hats to all officers whatsoever when they pass by them."

The History of the Royal Scots, by J. C. Leask and Captain McCance, quotes the following Regimental Standing Order issued by Major Dalrymple in 1762: "As nothing disfigures hats or dirties the lace more than taking off the hats, the men are for the future only to raise the back of their hands to them with a brisk motion when they pass an officer. When they have occasion to take off their hats entirely it must be with great care." Here is practically the modern form of salute, but whether the Royal Scots were the first to adopt it or it had already been practised in other regiments is uncertain. The officers, however, continued to take off their hats, as in *The Discipline of the Norfolk Militia*, first published 1759 and again in 1762, the illustrations show them performing the marching and standing salute with the fusee.

FIG. 181. Officer Saluting.

The Standing salute is described: "1st Motion: Seize the fusee with your left hand at the swell of the tail pipe, bringing it a hand's breadth from the shoulder and keeping it upright. 2nd Motion: Step back with the right foot a moderate pace or 18 inches, in line with your left heel, your right toe pointing to the right and the left to the front, keeping your body very upright; and drop the point of the bayonet directly to the front within 8 inches of the ground; supporting the piece on the back of the left hand, holding it slightly between the thumb and forefinger, the fingers extended, and back of the hand upwards; grasping the small of the stock with your right hand, the right elbow square, at the height of the shoulder. 3rd Motion: Bring your right foot up square, and raise up the fusee perpendicular, as in the first motion. 4th Motion: Bring your fusee on your right arm as in explanation of 2nd motion and put your left hand up to your hat, the left elbow square. 5th Motion: Pull off your hat with your left hand and let it hang down behind your sword, taking care not to bow your head in the hat" (Fig. 182).

FIG. 182. Officer's Standing Salute, with Fusee.

The Marching salute was performed in five motions: "1st, the positions are the same in the salute standing. It is to begin when you are at least about six paces from the person whom you are to salute; observing to begin the 1st motion when you step with your right foot. 2nd, the second with the left, which brings you to the second position. 3rd, with your right stepping forward with it, which answers to the 3rd position. The 4th, with the left, and 5th, with the right. After you are past the person whom you have saluted about six paces, put on your hat again in two motions; 1st, put on your hat; 2nd, throw your left hand down by your side."

The custom of marching with arms reversed at a funeral was already in practice in the sixteenth century, and later, about the end of the seventeenth century, an engraving shows cavalry riding with reversed swords.

Stow, in his Chronicles, describes a soldier's funeral in St. Paul's Churchyard c. 1573, his comrades following the body with trailing pikes and firing their matchlocks over his grave. The same ceremonial was observed at the funeral of Sir Philip Sidney. At the Duke of Marlborough's funeral, 1722, besides the minute guns and salvoes fired by the artillery, three volleys were fired by 2,000 Foot Guards. It was on the same occasion that the order later given as "Mourn arms" was first used, the orders then being "Reverse your arms" and "Rest on your arms reversed," "every soldier turning his musket with the muzzle to the ground and bending his head over in a melancholy posture."

Military mourning has already been mentioned, and the various orders for that to be worn at Royal funerals by officers state that "the scarlet coat is to be buttoned to the waist with a mourning button, facings of black cloth, black waistcoats and breeches, the sashes to be covered in *crêpe*. The hats to be without lace and *crêpe* hat bands with the ends appearing at each corner. Mourning swords and buckles." According to some orders the lace is to be removed from the coat.

The colours and drums were also covered in *crêpe*. In the print of the funeral procession of the Duke of Marlborough the men are also shown with *crêpe* on the hats and caps and the Pensioners of Chelsea Hospital are wearing long black coats or cloaks.

HANOVERIAN REGIMENTS

Fig. 183a. Foot Guard Regiment, 1758.

Fig. 183b. 1a Regt. 1758.

Although, strictly speaking, the Hanoverian regiments did not form part of the British army, the Georges being Kings of Great Britain and Hanover certainly regarded the armies of both countries as their own. This, perhaps, was particularly the case with George I and II, whose foreign policy was more especially conceived in the interest of the Electorate than in that of Great Britain.

The Hanoverian regiments formed part of the British armies on the Continent, and are to be found frequently brigaded with our own units.

For those interested in uniforms some general knowledge of these regiments is useful, as the similarity of their uniforms to the British and the use of the Royal coats of arms and badges of England have on occasion led to error in the identification of regiments.

It is impossible in this work to give more than a general outline of the uniform, even if space were available, as it is a study in itself. For those wishing to go more thoroughly into the subject Mr. Frederick Schirmer's *Nec Aspera Terrent* in 3 vols. and *Das Celler Soldaten Buch* will be found a mine of information, which the author has kindly allowed me to use.

Before going further let us take the sources of the information used.

In the British Museum are a number of contemporary sketches of soldiers, coats, waistcoats, horse furniture, etc., besides lists of Colonels and dates of their commands. The earliest periods shown by the drawings are 1729 and 1750 (Figs. 129 A and 187 A, B, D, and E, 211 and 212).

Then comes Morier's series of paintings of Hanoverians in Windsor Castle, similar to the British set, showing a grenadier of the infantry regiments and a few cavalry. All the infantry uniforms are similar in cut and style of lacing except in the colour of the facings, lace and the devices and colour of their cloth caps. Unfortunately no titles are given so identification is conjectural (Figs. 188, 208 and 209).

The next period is covered by a volume of contemporary drawings in the Celle Museum showing the uniforms of the army in 1758–62. For copies and sketches from this book and information from two works by Schutz von Brandes and von Sichart quoted in his *Nec Aspera Terrent* I am indebted to Mr. Schirmer. A small book of highly finished drawings of infantry regiments and copy of Raspe, both of 1758–60, in the King's Library, Windsor, have been an invaluable source of pictorial information. (Figs. 185, 197, 198, 200, 201). Besides the above R. Knotel in his *Uniformenkunde* gives pictures and information from various German sources.

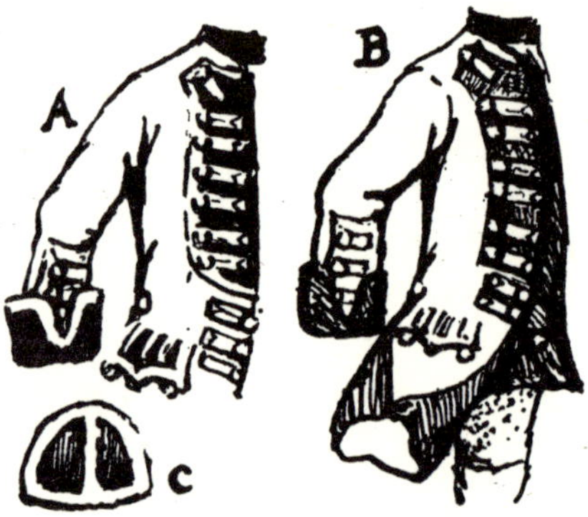

FIG. 184. Infantry coats.
A 1729. B 1750–1753. C Bandsmen's wing.

The Hanoverian infantry by 1742 consisted of some 21 regiments including the Foot Guards. These numbers were gradually increased to 27 by 1763. They were numbered by a system of letters and numerals—1 A, 2 A, 1 B, 2 B, and so on.

All the line regiments were dressed in red coats with half lapels, round cuffs and skirt turn-backs of the facing colour, with waistcoats to match. Both these garments were laced on the edges and buttonholes with a plain yellow or white worsted without any pattern or design. The Morier paintings and the British Museum 1750 MS. agree in these details. A red cloth strap was worn on the left shoulder of the coat. The cuffs of the officers and N.C.O.'s were furnished with a three-pointed slash and three buttons, but this was not given to the men until 1731. The buttons of the Foot Guards were gilt and bore the Royal arms of England, and those of the line had the

Horse of Hanover, which was white in the case of those regiments having brass buttons and yellow in those having theirs of white metal. The breeches were blue for the Guards (although Morier and Knötel give them yellow) and straw colour for the other regiments. It is mentioned in one place that for full dress red breeches were worn but there appears to be no pictorial evidence of this except in one instance, of an artilleryman. The straw colour is shown variously, either a buffish or bright yellow. The garters were white, and Morier shows all regiments with yellow garters. The Grenadiers wore mitre-shaped red cloth caps, with the front, little flap and turn-up generally of the facing colour. The fronts were decorated in various ways, sometimes the G R or white Horse within the Garter with a small L underneath, the whole surmounted by the crown, at others the Royal arms and supporters. In Morier's paintings these appear to be cut out metal plates of white or yellow. The little flap was embroidered round the edge with *Nec Aspera Terrent* and ornamented with the Horse, a trophy of arms or a grenade. They are also cut in more varied designs than the British rectangular oblong. The turn-up at the back was embroidered with a burning grenade in red. The cap of the Foot Guards is shown in one case with the Royal arms and supporters, and in Morier's paintings by the Garter, star and side scrolls, much like British Guards (Fig. 207, 1).

The officers' gorgets were gilt or silver to match the coat buttons and were fastened by ribbons of the facing colour or by a small chain, those of the Guards being decorated with the Royal arms and those of the line with the Horse.

Officers wore yellow sashes over the right shoulder, and carried swords with yellow and gold or silver knots. The Battalion officers were armed with spontoons, and the Grenadiers after 1732 had fuzils. Their uniforms were laced on the lapels and cuffs with gold or silver according to the colour of the buttons. The higher ranks of the N.C.O.'s also had theirs laced in the same way, but no doubt with a narrower lace. They carried swords having knots of white or yellow silk. The Sergeants were armed with halberts, those of the Grenadiers with muskets. The Corporals' uniforms were laced with white or yellow worsted and had sword knots of the same. The stocks were red at first and later black. The Privates were armed with flint locks, bayonets and swords. The Grenadiers carried grenades in a large pouch on a buff belt on which was fixed the match case. The pouches of the officers were white and

A — Saxe-Weimar Regt.
B — Invalids.
C — Major Appelbom's 3rd Frei Bn.
D — Major Usham's 4th Frei Bn.

FIG. 185.

decorated with a large grenade in the centre and a smaller one at each corner of the flap. Pioneers, or Charpentiers as they were termed, wore mitre caps like the Grenadiers, but with metal chains round the sides. The drummers and musicians had coats with "wings" on the shoulders and ornamented with lace (Fig. 184C).

The drums were of brass, and the hoops were usually painted the facing colour.

The Hanoverian soldiers wore moustaches, and the hair was worn with a more definite curl turned upwards than shown by Morier in his paintings of the English soldiers. The pigtail was worn by all except in the case of the Grenadiers, where it is shown plaited and turned up at the back of the head.

According to the pictures the lace was taken off the lapel edges and the waistcoats were left quite plain. A pom-pom, or ball tuft, of different colours was worn in the hat and two coloured tassels are seen at the side corners. The hats were bound with white or yellow lace, and ornamented with a black cockade. In some cases a green cockade is shown, especially by Frei Corps.

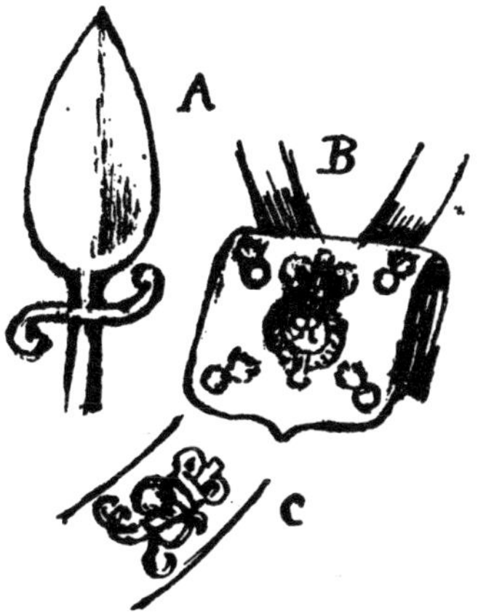

FIG. 186. Officers espontoon, Foot Guards pouch belt and Dragoon pouch.

After the Line come 32 Land companies, which were formed into seven battalions. Their dress was a red coat, white lapels, cuffs and turn-back; waistcoat and breeches straw colour. The lapels had three pairs of white metal buttons. One version of 1759 gives no lapels and red facings. Hats laced white with pompom and tassels to match.

The Garrison Battalions were three in number (in 1762). The coat was red, with red cuffs, lapels and waistcoats. The lapels had six white metal buttons in pairs and the turn-backs and breeches were straw coloured, hat laced white, pompom and tassels red and white. Black gaiters.

The Invalid companies according to the Windsor book of 1759 wore red coats without lapels, red cuffs, white skirts and waistcoats. In 1763 the uniform is described as red with blue lapels, cuffs, turn-backs and waistcoats. The blue was a much lighter tone than the usual dark blue. The lapels had three pairs of white buttonholes and buttons; hat lace white and the pompom and tassels red and white; breeches straw colour, gaiters and stocks black (Fig. 000). The Neustadt had green facings and white buttons; Clausthal straw colour; Bentheim Independent Company yellow with white laces and buttons, another version giving blue facings;

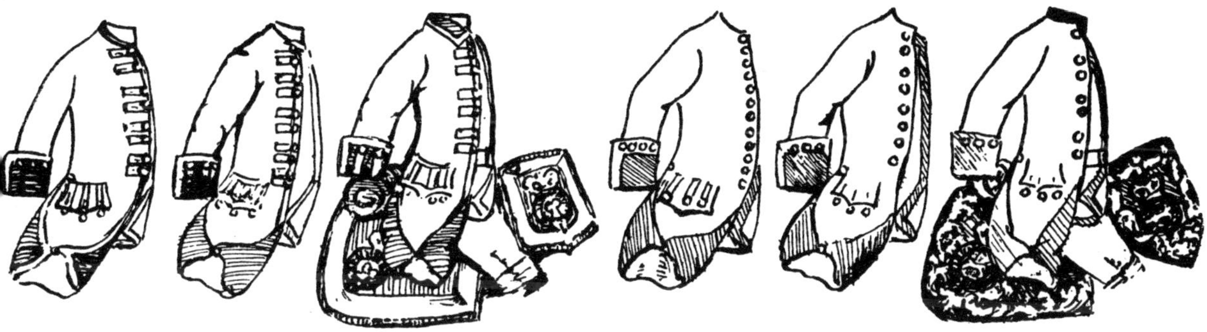

A 1729 B 1750 Garde du Corps C 1758 D 1729 E 1750 2A Cavalry Regiment F 1758

FIG. 187.

FIG. 188. 1A Leib Regiment.

Hoher Schanze B^{n} green with white metal buttons; Burgschanze red facings, yellow lace, brass buttons; Scharzfeld bright blue, straw coloured turn-backs and white metal buttons.

The Cavalry consisted of eight regiments of Horse or Reiter Regiments, and four of Dragoons besides the Leib Garde or Household Troops. The regiments of Horse were numbered on the same system as the Foot from 1 A to 4 B. After these came the four Dragoon regiments numbered 5th to 8th without any accompanying letter.

The uniform of the Cavalry in 1729 was a white coat (without lapels), round cuffs and skirt turn-backs of the facing colour. One German authority states that the coat had a small turn-down collar of the facing, but another writer says that this was by no means usual. The British Museum MS. shows only one regiment, 4 B, with a blue collar. According to one list for 1729–30

2 B, 3 A and 4 B are given lapels. The British Museum MS. shows none for the Horse and only one, 3 A, has a red shoulder knot. It gives a large number of buttons on the front of the coats varying from 8 to 14.

The hats of both Horse and Dragoons were bound with gold or silver coloured lace and ornamented with the black cockade and a sprig of oak leaves.

The stocks were black except for the Horse Grenadiers and 1 B regiment, who had red ones. The 8th Dragoons are also given red stocks in one case.

The hat about 1758 appears to have been worn with the front corner more raised than the sides, making it less three-cornered in form than the British.

Dragoons also wore white coats with lapels, cuffs and skirts of the facing. The lapels were laced on the edges and buttonholes and a shoulder knot of facing colour terminated by tassels of the colour of the lace. The British Museum MS., however, shows metal points instead of tassels.

The Dragoons had Troops of Grenadiers, who wore mitre caps with metal plates. The earlier caps had white crowns and fronts of the facing colour with the white horse on a red scroll. The Grenadiers of the 6th regiment had bearskin caps.

In 1741 the fronts of the caps of the 5th and 6th Dragoons were white and blue respectively.

The breeches and waistcoat of both Horse and Dragoons were straw coloured, and the waistcoats were usually bound on the edges with the facing cloth and were fastened back like the coat skirts.

The housings and holster caps were of the facing colour, and usually most elaborately ornamented with complicated scroll designs and lace edgings. Some of the holster caps instead of the usual square pattern were of a more pointed form resembling a shield. The cypher G R or the white horse within the Garter surmounted by the crown, usually with a scroll design either side, was worked on the corner of the housings and holster caps. The 5th Dragoons had the white horse within a cartouche of black and white scroll pattern and no Garter.

HORSE FURNITURE FOR CAVALRY

1 A. *Leib Regiment.* Yellow housings with a red border with black half circles, blue, white and red scroll pattern with a few black spots (Figs. 188 and 189).

1A Leib Regiment. 1B

FIG. 189. Housings and Holster Caps of Cavalry, 1758.

1 B. Orange housings with two bands of light blue with double zig-zag of yellow (Fig. 189).

2 A. Light green housings with yellow, white and red scroll pattern. The holster caps have, in addition to this, red and yellow circles and spots (Fig. 187 F).

2 B. Dark green housings with a border of yellow and white oblongs edged round with red. Red, yellow and white scrolls. Holster caps have at the top a white lattice design edged red and red and white scrolls, and a yellow strip issuing from a red and white device in the corners (Fig. 190).

3 A. Crimson red housings with a border of a white spiral scroll edged yellow between two yellow stripes, yellow and white scrolls or foliage either side of the Crown and Garter (Fig. 191).

2 B 3 B

FIG. 190. Housings and Holster Caps of Cavalry, 1758.

FIG. 191. 3A Cavalry Regiment, 1758.
Detail of Housing.

FIG. 192. 7th Dragoon Regiment.
Grenadier Cap and Housing, *c.* 1758.

4 A 4 B

FIG. 193, Housings and Holster Caps of Cavalry, 1758.

3 B. Scarlet housings with a border of three stripes of red and black diagonals edged yellow, white and yellow scroll pattern (Fig. 190).

4 A. Dark blue housings edged with white and yellow diagonals. Red, white and yellow design of foliage (Fig. 193).

4 B. Sky blue housings with border of a wide band of red edged yellow and on either side, separated by blue, two narrow stripes of yellow (Fig. 193).

DRAGOONS

Fifth Dragoons. Red housings with a border of same colour edged white and having a black zig-zag edged white entwining a central thread or line of yellow. Black and white scroll design (Fig. 195).

6th Dragoons. Light blue with narrow outer edge of red on white and a broader one of two red stripes on white. White scrolls (Fig. 194).

6th Dragoon Regiment. 8th Dragoon Regiment.

FIG. 194. Housings and Holster Caps of Dragoons, 1758.

7th Dragoons. Bright blue housings with outer border of white edged yellow with blue zig-zag, an inner band of red edged yellow with a white scroll entwining a white central stripe: white scroll pattern (Fig. 000).

8th Dragoons. Scarlet housings with a border of yellow edged with blue and having red diamonds with blue centres as the central pattern, on either side of this is a white triangular design on red edged with white. The holster caps have the outer edging of the white triangles. The inner pattern consists of a yellow stripe forming a triangular design, and a yellow triangle fills each corner of the holster cap; on each side is another triangle of white with a blue centre; a second white stripe passes across the corners of the holster and forms a triangle on each side of the central panel; a yellow zig-zag forms the inner pattern. All these different stripes are separated by the red cloth

FIG. 195. 5th Dragoon Regiment. Grenadier Troop and Housing, 1758.

of the holster cap. The central device of all these housings is the white horse or the G R on a red ground within the Garter and surmounted by the crown (Fig. 194).

The decorations on these housings may appear to be too complicated to have really existed, but at the Musée de l'Armée, in Les Invalides, is, or was, an actual housing of red cloth with a design almost as elaborate with the G R cypher and crown. It is either British or Hanoverian of this period.

The Cavalry were armed with swords, a pair of pistols and a carbine, and Dragoons with swords, pistols and muskets, with short bayonets. These last were carried in 1730 on a shoulder belt.

The Leib Guard consisted of the Garde du Corps numbered LG.A, and the Horse Grenadiers LG.B.

The Garde du Corps wore red coats without lapels, blue cuffs and turn-backs. The front of the coat had three pairs of silver laced buttonholes. The cuffs had two rows of lace and the pocket flaps were edged with silver. The British Museum MS. sketches for the years 1729 and 1750 both agree with these details. Other versions show blue lapels and collars laced with silver and one other gives red cuffs. The hat lace was silver and the shoulder belts were buff with a central stripe of blue and outer edges laced with silver. The cloak was blue. The housings and holster caps were red edged with silver and had the G R within the Garter surmounted by the crown worked in silver. The corps was mounted on grey horses (Fig. 187 A, B, C).

The Horse Grenadiers had red coats, red skirt turn-backs, black lapels, and cuffs edged gold. Knötel's illustration shows no lace on the buttonholes, while the British Museum MS. gives gold loops on lapels and cuffs. A red and gold shoulder knot was worn on the right shoulder.

The cloth cap was red with black velvet front embroidered with the Royal arms and supporters, the little flap being red edged gold and bearing the white horse. The tuft was red and gold (Fig. 196).

The British Museum MS. has a drawing of a Horse Grenadier with slight variations to the above, having the G R on the front of the cap and laced buttonholes on lapels and cuffs. The horse furniture was red with a zigzag lace of gold with a black centre and another on the outer edge of gold with black diamond pattern running through it. On the housings and holster caps were the cypher G R and crown in gold. The British Museum MS. shows red housings with gold diagonal stripes between two gold laces and on the inner side is a wide scalloped edging. The G R is encircled by the Garter. In place of the rolled red cloak of Knötel's illustration is an all red valise. A later version by Raspe shows those parts of the uniform blue that were formerly black, and silver instead of gold lace. The cap is red with a blue front and turn-up. A contemporary water-colour in the Bibliothèque Nationale agrees with this colouring. The equipment was buff and the pouch black, Knötel showing a red edging and brass plaque with G R in the centre with a small grenade in each corner. The sword belts were brass with a shell design. The

FIG. 196. Horse Grenadier, housing, and sword hilt.

girths were red and white. R. Knötel shows a white watering bridle and bridoon and the British Museum MS. red bridoon rein. The men were mounted on black or dark brown horses (Fig. 196).

One of Morier's paintings shows the back view of a Horse Grenadier in marching order. He wears a red coat faced with blue and a fur cap with the hood or bag hanging down behind. The housings are red with white lace with a blue zig-zag. Unfortunately the details do not correspond with anything that is known. We have seen that only cloth caps appear in the contemporary pictures and descriptions with the exception of those of the 6th Dragoons; when fur caps were eventually adopted by the Hanoverian army it was some years after the Horse Grenadiers had been disembodied. Whether this figure represents a head-dress worn previously to 1729 or was only a project, remains for the German experts to determine.

Besides the regular regiments, various Free Corps of light troops were raised during the Seven Years' War. Schülenburg's Jäger Corps formed in 1757 became Freitags in 1759, and was composed of mounted and infantry companies. The uniform was green coats with lapels, cuffs, linings and waistcoats of the same colour with white metal buttons, white breeches and grey gaiters. The hat was plain with a black cockade, green rosette and loop and a sprig of oak leaves. A later version gives brown waist belts and buff gaiters. The mounted men wore the same uniform with black boots and buff equipment and horse furniture of goat or sheep skins. Raspe shows a mounted Jäger with all green coat and lapels, green waistcoat, breeches and cloak. No horse furniture is shown except the green valise edged white and a white striped blanket under the saddle. This may represent the dress when amalgamated with Freitags (Fig. 198).

FIG. 197.. Grenadier, Stockhausen's Frei Corps.

Major Stockhausen's Light Troops were raised in 1759. They consisted of

FIG. 198. Mounted Jäger, Freitag's Frei Corps.

one company of Grenadiers and two of Jägers. The uniform was a green coat with green lapels, cuffs, lining, waistcoats and breeches with white metal buttons on cuffs and lapels and short-pointed light infantry gaiters.

The Grenadiers had low caps entirely of green with metal mountings like the Prussians. On the front were the arms of England on a shield with the crown and supporters in white metal. The little flap was edged with white and had the white horse, on the turn-up at the back were white metal grenades (Figs. 197 and 199).

The musket slings were black and a brown waist belt carried a black pouch as well as a sword and bayonet.

The mounted squadron, raised 1760, wore the same dress as the Jägers but with straw coloured breeches and boots, the head-dress being the cocked hat with oak leaves.

Fig. 199. Stockhausen's Frei Corps. Grenadier Cap.

In 1761 Stockhausen's and Freitag's Corps were amalgamated. R. Knötel shows the uniform much as described above, but the infantry differ in having white breeches and high boots.

In 1758 a Corps was raised by Captain Scheither, originally as a Corps of Grenadiers and by 1761 there were both mounted and foot companies. The Carabiniers, as the mounted section was called, are shown dressed in white coats bound with green cloth down the front and round the skirt turnback, green turn-down collar edged white and round green cuffs. The waistcoat was green edged white, straw or buff breeches and high boots. The hat was plain black with cockade and loop. In some copies of Raspe the cockade is shown green. The shabrack was green edged with a white lace having black zig-zag lines running through it, the crown and horse in white on the rear. The front is hidden by the green cloak which is loosely strapped over the saddle. The valise was green edged white on the ends. The sabretache was green with G R, crown and edging in white (Fig. 200).

Fig. 200. Carabineer, Scheither's Frei Corps.

The Grenadiers wore green coats with green cuffs and linings, the buttons of white metal, the pockets were bound with a mixed white and yellow lace, waistcoats and breeches

Grenadier. Carpenter. Artillery.
FIG. 201. Scheither's Frei Corps.

straw coloured. A brown fur cap without any plate, which, according to Schirmer, had also a green cord hanging down at the back (Figs. 201 & 203).

The "Carpenter" wears the same dress as the Grenadier, but has a small black helmet with metal crest and a green band or turban with a white pattern. On the front is the white horse on a green ground, a small crimson plume and a green rosette are worn on the left side. The gaiters are light brown instead of white like the Grenadiers (Figs. 201 and 202).

The Luckner Hussars were formed in 1759 from men in the Dutch service. Their first dress was a dark green dolman braided with yellow, pelisse of the same colour with brown fur and red breeches, a "flugel" cap (i.e. one resembling the high watering cap or the mirleton of the French Hussars of later date) with yellow lace, yellow Hessian boots with red and yellow tassels. The sabretache was red edged yellow and was ornamented with the crown and white horse. It was worn on very short slings so that its upper edge was almost touching the bottom of the dolman. The cloak was green. The horse furniture was brown sheep or goat skin with red border (Fig. 204).

The next uniform was a grey fur Hussar cap with a small black feather, red bag and white cap lines; white dolman with yellow braid and cuffs; red pelisse, brass buttons and yellow braid and black fur. The breeches were white and ornamented with lace down the front of the thighs. A yellow girdle was worn round the waist. The sabretache was red edged yellow, with the white horse on a green ground. Housings were edged red and bearing the

FIG. 202. Carpenter's Cap, Scheither's Frei Corps.

white horse. The valise was red edged with yellow and the cloak white. The equipment was buff or straw colour (Fig. 205).

A Corps entitled the Légion Britannique was raised during the Seven Years' War. It consisted of five battalions of infantry to each of which was attached a squadron of Dragoons. These were later formed into one regiment. The Dragoons wore the same dress. The housings were red edged with yellow lace and had rounded corners.

The Legion seems to have been recruited from other nationalities besides Hanoverians, as Hessians, Westphalians, Brunswicks, Dutch and even a Russian appear in the ranks and the lists of officers show a sprinkling of British and French names. The Hanoverian regular regiments supplied detachments of officers, N.C.O.'s and men.

FIG. 203. Grenadier, Scheither's Frei Corps.

Uniform:

1st B^{n} Light blue coats, brass buttons, straw coloured facings, linings, waistcoats and breeches.

2nd B^{n} Medium blue coats, white metal buttons, scarlet facings and linings, white waistcoats and straw coloured breeches.

3rd B^{n} White coats, white metal buttons, orange facings, linings and waistcoats, straw coloured breeches.

4th B^{n} Red coats, white metal buttons, bright blue facings, linings and waistcoats, straw coloured breeches.

5th B^{n} Red coats, white metal buttons, black facings, white lining and waistcoats, straw coloured breeches (Fig. 206).

FIG. 204. Luckner Hussars.

FIG. 205. Luckner Hussars.

All battalions had round cuffs without any slash; equipment buff and pouches black.

The coats of the Dragoons were without lapels and the hats were plain with a light green cockade, the tassels of the 1st being gold and the rest silver. The shoulder knots of the 1st were yellow, the 3rd orange, and the others silver grey. The equipment was buff.

The housings of the 2nd, 3rd, and 4th were red edged with lace. The G R and crown were on the housings and holster caps of all the squadrons. The girths were red and white for the 1st, 4th and 5th white and blue, the 2nd white and orange, the 3rd.

FIG. 206. Legion Britannique, 5th Bn.

The cloaks and valises were the colour of the facings of the battalion. The valises were edged with white lace. The bridoons were red.

In the following list of Morier paintings the pattern of lacing on lapels and coats is the same. The waistcoat, except where mentioned, is of the facing colour, and is always laced in the same way for all regiments. The Royal Arms are emblazoned on a shield, and in some cases the whole coat of arms appears to be placed on a cartouche. They are generally in full colours except when they are shown all in yellow or white they appear to be on an embossed metal plate, which in some cases seems to cover the place of the "little flap." This last is generally of a more cut-out shape to the usual English shape, and the embroidery on it is often of a very fine sprig letter design. The cap is nearly always shown front view, but in those instances where the side can be seen the turn up is the facing colour and the crown red.

Foot Guards. Blue facings, yellow lace, cap blue with Garter Star with crown above, white foliage either side; little flap red with the White Horse. Tuft gold (Fig. 207[1]).

1. Blue facings, white lace, a difference to other regiments being that the three loops below the lapels are diamond-shaped, the rest being the usual pattern. Cap, blue; G R in yellow on red within the Garter,

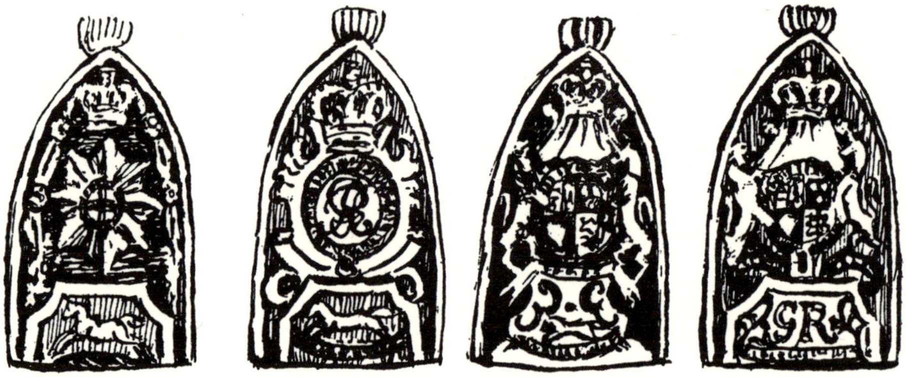

1 2 3 4

FIG. 207. Grenadier's Caps.

Crown above, white scrolls either side, little flap red with white grenade and scrolls.

2. Blue facings, straw coloured waistcoat, yellow lace. Cap red edged white, the Royal Arms within the garter, supporters, crown above, scroll below with *Dieu et mon droit*. The whole on a white metal plate; little flap blue edged white with G R and scrolls in white; tuft white. (Fig. 208).

3. Blue facings; yellow lace; cap, metal front with the Royal Arms within the garter. Supporters and crown. Little flap, a trophy of arms in yellow.

4. Black facings, straw coloured waistcoat, white lace. Cap black, G R in yellow on red within the garter, crown above, white scrolls and foliage on either side. Little flap with grenade and scrolls: red and white tuft.

FIG. 208.
Grenadier, 1750.

5. Black facings, straw coloured waistcoat, yellow lace. Cap, black with the Royal Arms within the garter; supporters; crown above, below a ribbon with *Dieu et mon droit;* beneath this again the Horse and scrolls and label with *Nec aspera terrent*, the whole forming a brass plate. Tuft, yellow and black (Fig. 207[3]).

6. Light blue facings, white lace; cap, light blue, the White Horse on red within the garter, crown above, with a surround of white scrolls. Little flap, red edged white, with white grenade and scrolls. Tuft, blue and white.

7. Deep orange facings, yellow lace; cap orange, brass plate with Royal Arms, garter, supporters and crown; no flap is shown, its place being covered by a scroll design. Tuft, red (Fig. 210[1]).

8. Facings and lace as above; cap orange with White Horse on red within the garter; Royal supporters and crown; below a blue ribbon with *Nec aspera terrent*. The little flap has a grenade between a flag and a guidon, and scrolls. (Fig. 210[4]).

9. Facings and lace as before; cap orange with Royal Arms, supporters and crown, embroidered in

gold colour. Little flap orange with the Horse and scrolls. Tuft, white (Fig. 210 [2]).

10. Dull yellow ochre facings, white lace; cap red; on an orange cartouche, G R on red within the garter, crown above. Little flap red edged white, the White Horse with blue label below with *Nec aspera terrent*. Tuft, red and white (Fig. 207 [2]).

11. Yellow facings, yellow lace; cap yellow, the White Horse on red within the garter and crown over; the whole on an orange cartouche, and surrounded with white scrolls and foliage. The little flap yellow, edged white, with white grenade and scrolls (Fig. 210 [3]).

12. Pale yellow facings, white lace; cap yellow, with white metal plate with Royal Arms within the garter, supporters and crown. Little flap, yellow, G R in centre of trophy in white. Tuft, white.

13. Facings and lace as before; cap yellow with White Horse on red within the garter, crown above; on either side white scrolls. Little flap, red edged white, with a white grenade.

FIG. 209. Grenadier.

14. Olive green facings, white lace; cap green, G R in white on red within the garter, crown above, white and yellow scroll design round. Little flap green, edged white, a white grenade with two smaller grenades in red and yellow and white sprigs either side (Fig. 209).

15. White facings, yellow lace; cap, white, edged yellow. The White Horse on red within the garter; Royal supporters and crown: below a blue ribbon with *Nec aspera terrent*. Little flap, white, edged yellow, with trophy of arms in yellow. Tuft, red and white.

16. White facings and lace; cap, white, on a brass plate the Royal Arms within the garter and crown above; the whole on a cartouche of scroll design. The little flap white, edged yellow with G R and scrolls. Tuft, red and white.

17. Facings and lace as before; cap white with white metal plate with

Royal Arms, supporters and crown. On the little flap G R and scrolls. Tuft, white.

18. White facings; yellow lace; cap white, edged yellow. The White Horse within the garter, crown and Royal supporters and yellow scroll with *Dieu et mon droit*, in little flap red with Trophy of Arms in yellow (Fig. 210 [5]).
19. Light blue facings, yellow lace, white waistcoat laced yellow, Red cap Royal Arms within the garter; Royal supporters. The whole forming a white metal plate (?). The little flap blue with G R sprigs and below a ribbon with *Nec aspera terrent* (Fig. 207 [4]).
20. White facings and lace. The whole front of the cap appears to be covered by a white metal plate the little flap included; this indicated by raised metal lines. The Royal Arms within the garter supporters a crown; at either side of the base often are a Rose and Thistle.
    Below the garter is another and smaller crown. The little flap has G R in cypher with ornamental repousé work either side.

The above numbers have no reference to those of the regiments as it is not easy to identify most of them, but a reference to the list may help to do so.

(*See over.*)

1 2 3 4 5

FIG. 210. Grenadier Caps.

INFANTRY

Foot Guards, 1731, facings dark blue, straw coloured waistcoats.
1742, blue waistcoats, gold lace and buttons, Pompoms.
1748, blue and red.
1756, blue tassels.
1 A. 1724, dark green facings, straw coloured linings and waistcoats, yellow lace and buttons.
1742, same, but dark green linings and waistcoats.
1748, red and green Pompoms.
1756, green and yellow Pompoms and tassels.
M.B. 1724 and 1750 same details.
1 B. 1731, white facings and waistcoat, yellow lace and buttons.
1748, Pompoms red and white.
1756, red and yellow Pompoms and tassels.
B.M. 1724 and 1750 same details.
2 A. 1731, facings straw, yellow lace and buttons.
1742, Pompom straw and red.
1756, red and yellow Pompom and tassels.
B.M. 1729 same details, 1750 dark buff facings, etc.
2 B. 1717, facings yellow, silver lace and buttons.
1748, straw and red Pompom.
1756, straw tassels.
B.M. 1729, yellow lace; 1750 agree with first details.
3 A. 1731, black facings, straw waistcoats, silver lace and buttons.
1742, white linings and waistcoats.
1748, red and white Pompom.
1756, red, black and white Pompom and red and white tassels.
B.M. 1729 and 1750 give same details.
3 B. 1731, black facings, yellow lining and waistcoat, yellow lace and buttons.
1742 and 1748, same details but with yellow and red Pompom.
1756, yellow, black tassels.
B.M. 1729 same, but silver lace; 1750 details as first lists.

4 A. 1741, red facings.
1748, medium blue facings and linings, blue and red Pompom, silver lace and buttons.
1756, red, blue and white Pompoms and blue tassels.
B.M. 1729, 1750 same details.

4 B. 1717, black facings and linings, gold lace and buttons.
1742, red linings and straw waistcoats.
1748, same with red and straw Pompoms.
1756, yellow and white Pompoms and yellow tassels.
B.M. lists the same details.

5 A. 1724, lemon facings, white linings and waistcoats, silver lace and buttons.
1742, yellow facings, lining and waistcoats.
1748, bright orange facings, lining and waistcoats, Pompom orange and red.
1756, yellow facings, etc., red, yellow and white Pompom and yellow tassels.
B.M. 1729 yellow facings, 1750 deep yellow facings, etc.

5 B. 1724, red facings.
1742, yellow facings, lining and waistcoat, gold lace and buttons.
1748, orange facings, etc., red and orange Pompom.
1756, straw facings, etc., yellow Pompom and tassels, yellow lace.
B.M. 1729 red lapels, straw cuffs, silver lace.
1750, orange facings, etc., gold lace and buttons.

6 A. 1725, facings red, linings blue.
1730, orange facings, white lining and waistcoat, silver lace.
1742, yellow facings, linings and waistcoats.
1748, orange facings, etc., orange and red Pompom.
1756, red and yellow Pompom and tassels.
B.M. 1729 red facings, silver lace.
1750, straw facings, etc.

6 B. 1730, lemon facings, linings and waistcoats, silver lace and buttons.
1742, the same.

1748, orange yellow facings, etc., red and orange Pompom.
1756, yellow facings, etc., red, yellow and white Pompom, red tassels.
B.M. red facings, gold lace. 1750, orange facings, etc., silver lace.

7 A. 1730, straw facings, linings and waistcoats, silver lace.
1742, the same.
1748, the same.
1756, straw Pompom and tassels.
B.M. 1729 orange facings, 1750 yellow facings, silver lace.

7 B. B.M. 1729, straw facings and lining, gold lace and buttons, 1742 the same, 1748 straw and red Pompom, 1756 yellow facings, etc., straw tassels.
B.M. 1750 same details.

8 A. 1729 (B.M.) red lapels, buff cuffs.
1742, white facings, lining and waistcoat, silver lace and buttons.
1748, red and white Pompom.
1756, white tassels.
B.M. 1750, same details.

8 B. 1729 (B.M.) red lapels, white cuffs, silver lace.
1742, the same.
1748, white lapels, lining and waistcoat, red and white Pompom
1756, red and white tassels.
B.M. 1750 same details.

9A. 1756, green facings, linings and waistcoats, green and red Pompom, silver lace and buttons.

9 B. 1718, dark green facings, linings and waistcoats, silver lace and buttons.
1748, red and green Pompom.
1756, same as above with green and white Pompom and green tassels.
B.M. 1729 and 1750 same details.

10 A. 1729 (B.M.) dark green facings, silver lace.
1741, green facings, white lining and waistcoat, silver lace.
1748, the same, red and green Pompom.

1756, the same with red, green and white Pompom and green and white tassels.

B.M. 1750, same details.

11 A. 1730, yellow facings, lining and waistcoat straw, lace and buttons gold.

1742, the same.

1748, red and yellow Pompom.

1756, the same.

B.M. 1729 and 1750, same details.

11 B. 1730, yellow facings, lining and waistcoat, gold lace and buttons.

1742, the same.

1748, yellow and red Pompom.

1756, the same.

B.M. 1729 the same but with silver lace; 1750, the same with gold lace.

12 A. 1748, light green facings, lining and waistcoat, silver lace and buttons, red and green Pompom.

1756, the same with green and white Pompom.

B.M. the same details.

12 B. 1742, red facings and wiastcoat, white lining (a note says it had no lace).

1748, silver lace and buttons, red and green Pompom.

1756, green and white Pompom.

B.M. 1750, same details.

13 A. 1750 (B.M.), blue facings, lining and waistcoat, silver lace and buttons.

1756, the same with red and blue Pompom and tassels.

13 B. red, red facings, white lining and waistcoat, white and red pompom.

CAVALRY

1 A. Leib Regiment 1729 (B.M.), orange facings and gold buttons.
1730, yellow facings.
1748 and 1750 (B.M.), the same with silver buttons.
1756, gold buttons.

1 B. 1729 (B.M.), 1745, 1748, 1750, orange facings, gold buttons, red pompom and stocks.

2 A. 1729 and B.M., 1729, 1748 (B.M.), 1750 light green facings and gold buttons.

2 B. 1729, dark green facings and lapels.
B.M., same without lapels, gold buttons.
1748 and B.M. 1750, the same.

3 A. 1729 (B.M.) and 1730, red facings and silver buttons.
1748 and 1756 the same, with gold buttons, green and white Pompom.

3 B. 1729 (B.M.), red facings and shoulder knot, silver buttons, 1730, 1748, 1750 and 1756, same details.

4 A. 1729 (B.M.), blue facings, gold buttons.
1729–1750 (B.M.), the same.
1756, white Pompom.

4 B. 1729–17 (B.M.), light blue facings, silver buttons, 1748, 1750 and 1756, same with gold buttons.

5th Dragoons—
1725–1729 (B.M.) 1748, 1750 (B.M.) and 1756, red facings and silver buttons.

6th Dragoons—all lists give light blue facings and silver lace.

7th Dragoons—all lists give bright blue facings and gold lace.

8th Dragoons—all lists give red facings, gold buttons and lace.
The B.M. lists all show lapels with lace edging and loops.

THE uniform of the artillery in 1748 was a light grey blue coat with red lapels, cuffs and linings and brass buttons. The hat was laced with gold, and had a black cockade with gold loop.

The MS. at the British Museum gives the sketch of a coat and waistcoat entitled "Regiment of Artillery (Cellishe) 1750." The coat is grey blue with plain red cuffs, lapels and skirts, blue waistcoat and blue cuff slash and brass buttons (Fig. 211).

Officers of artillery carried espontoons with red staves in 1727.

In 1756 the uniform was light blue coat and waistcoat, brass buttons, red cuffs with blue slash, red lapels and skirts, straw coloured breeches, white gaiters. The 1758 book at Windsor shows a grey blue coat with red lapels and cuffs with a blue slash, red skirts, straw coloured breeches, white gaiters, black hat and cockade with gold loop. The equipment is buff with a pouch on a waist belt, the former having an oval brass plate. A powder flask is carried under the right arm.

The drivers wore red coats with bright blue cuffs and lapels, the latter having eight brass buttons in pairs. The skirt turn backs were red and the waistcoat light yellow. The hat was plain except for black tassels and the cockade with a brass button. The breeches were buff leather and heavy boots completed the dress.

The Engineer officers about 1748 wore grey blue coats with red lapels, cuffs and waistcoats, brass buttons. The hat was edged with gold lace and tassels and cockade loop of the same. The British Museum MS. drawing of the uniform, 1750, is a blue coat, red lapels and cuffs edged and looped with gold, red waistcoat with gold lace edging to front, pocket flaps and buttonholes.

FIG. 211. Gunner, *c.* 1750.

The Pontooniers in 1750–63 wore light blue coats with red cuffs and linings, no lapels, but 14 buttons in pairs down the front. The coats had red turn-down collars. The waistcoats were blue, breeches straw colour and gaiters black. The equipment was buff.

The Engineer officers of the same period wore light blue coats with red lapels and cuffs. The lapels had 6 gold loops and the cuff slash was laced with gold. The waistcoats and breeches were red and the former were laced with gold. A red and gold shoulder knot was worn and also a yellow waist belt over the waistcoat. High boots were worn, and in one picture white gaithers are shown (Fig. 212).

FIG. 212. Engineer Officer's Coat and Waistcoat.

The gun carriages and ammunition wagons of the train were painted red. Two guns drawn by three horses were also attached to each battalion besides the two tumbrils and the necessary drivers for the teams.

A picture in Windsor Castle entitled "Major Scheiter's Artillery" shows the dress of the gun crews attached to his Frei Corps. It consists of a light grey blue coat with red skirt linings and blue cuffs with a yellow or gold zig-zag lace, a blue waistcoat and buff or straw coloured breeches and brown leggings. The hat is without lace and has a green cockade. The shoulder and waist belts are buff as well as the pouch, which has an oval brass plate.

Each Infantry Battalion had two flags. With the exception of two battalions, the King's Colour was white, and bore on one side the Arms of England with the lion and unicorn and mottos *Hony soit qui mal y pense* and *Dieu et mon droit;* the entwined initials "G R" beneath a crown in the corners; on the reverse side the white rose on a red ground within garter, and the motto *Nec aspera terrent* (Fig. 213). The second colours all bore fanciful allegorical designs, characteristic of the period. Appended are brief descriptions of these, according to v. Wissel.

1A. SCHEITHER: Green. An upright shield with design of sword and buckler surrounded by clouds; the whole within a laurel wreath. A lion passant embroidered in gold. Motto above: *Ne cede malis* (According to Knötel the motto is placed below.)

1B. ALT-ZASTROW: Green. Lion couchant. Motto below: *Vigilantia vincit.*

2A. SPORCKEN: Pale straw colour. Mailed arm issuing from a cloud, holding a drawn sword bearing the word "Victoria." Motto above: *Si deus pro nobis quis contra nos.*

FIG. 213. 1st Colour, 3A Regiment.

2B. FABRICE: Probably pale straw colour. A lion holding a sword. Motto above: *Nec temere nec timide* (According to Knötel the motto was below.)

3A. KNESEBECK: Red. A crag with arrows pointing towards it; broken arrows at the foot of the crag. Motto below: *Excipit et frangit.*

3B. DRUCHTLEBEN: Black. Sampson wrenching the Lion's jaws apart. Motto above: *Non mortem sed infamiam vereor.*

4A. LEDEBOUR: Pale blue. Mailed arm issuing from a cloud, within a laurel wreath. Motto above: *Dulce et decorum est pro patria mori.*

4B. STOLZENBERG: Pale blue. Pallas Athene crowning a champion with a laurel wreath. Motto below: *Bellici virtutis præmium.*

5A. GROTE: Yellow. Lion holding a sword and shield. Motto above: *Haud animo trepidus.*

5B. BEHR: Yellow. Youth in armour between two marble pillars connected by a chain on which hang two crowns and three laurel wreaths at which he is striking with the sword. Motto above *Antiquæ avidus gloriæ*.

FIG. 214. Colour, 6A Regiment.

6A. HARDENBERG: Orange. A lion rampant armed with a sword above a trophy of arms. Motto above: *Offendo et defendo* (Fig. 214).

6B. ZANDRÉ: Yellow. Mars enthroned with a spirit flying. Motto above: *Nulli sine morte triumphi* (Fig. 215).

7A. WANGENHEIM: Pale straw colour. Female figure of Fame with scroll bearing the words *Extendere factis*.

7B. HAUSS: Pale straw colour. Pyramid charged with a serpent. Motto above: *Per ardua virtus*.

8A. DIEPENBROICK: King's colour white. Upon a trophy of arms a lion couchant embroidered in gold. Motto above: *Ut alii dormiant*.
Regimental colour, white; Roman warrior, right hand drawing a sword, left-hand on his heart; a plinth behind bearing crown, sceptre and sword. Motto above: *Regi et patriæ*.

FIG. 215. Colour, 6B Regiment.

8B. BLOCK: White. Mailed arm with sword issuing from a grey cloud, the hilt entwined with a laurel wreath. Motto above: *Re non verbis*.

9A. SACHS-GOTHA: Since 1759 King's colour similar to other regiments. Regimental colour green. Lion passant holding sword. Motto above: *Pro fide, rege et lege*.

9B. JUNG-ZASTROW: Dark green. Issuing from a cloud a mailed arm holding scroll inscribed with the name Jehova. Motto below: *Quis contra nos.*

10A. POST: 2 pale green flags. Mailed arm with sword issuing from a cloud, the whole within a laurel wreath. Motto above: *Cui vult.* (According to Knötel the motto is *Re non verbis*) (Fig. 216).

10B. 1. New Battalion. Light red. Pyramid surrounded by trophy of arms and laurel wreaths. Motto above: *Præmia futura.*

FIG. 216, Colour, 10A Regiment.

11A. DREVES: Yellow. Sun in front of globe, with signs of the zodiac. Motto below: *Non retrorsum.* (According to Knötel *Nunquam retrorsum.*)

11B. OBERG: Blue. Mailed arm bearing laurel entwined sword issuing from a cloud. Motto above: *Idem pacis medius qui belli.* (The correct wording should be *Idem pacis mediumque belli.*)

12A. KIELMANNSEGGE: Green. Warrior in armour with sword. Motto above: *Miscetur decori virtus.*

12B. BRUNCK: Green. Rocks emerging from the sea, surrounded by a device of clouds, lightning and gales. Motto above: *Tu ne cede malis* (Fig. 217).

FIG. 217. Colour, 12B Regiment.

13A. FUSILIER: Probably blue. 4 green wreaths and crowns. Motto above: *Præmia perata.*

Particulars of the flags as given in the Ancienneté List of 1756 differ in the following respects from those given by Wissel:

2B. Pale straw colour. A trophy of Arms. White motto band: *Nec temere nec timide.*

3A. Red. A shield with representation of a knight before a castle from which rockets are ascending. A white motto band above with the motto: *Impavidium ferient ruine.*

7A. As described by Wissel, but with a trophy of arms below the figure of Fame, and G R surmounted by a crown in the corners (Fig. 218).

Fig. 218. Colour, 7A Regiment.

7B. As described by Wissel, but the colour of the pyramid is given as red and the serpent blue. In the corners bursting grenades (Fig. 219).

9B. A scroll with the word Jehova, backed by a cloud. Behind the cloud flags and trophy of arms. Below a white motto band with the motto: *Quis contra nos.*

Fig. 219. Colour, 7B Regiment.

11A. Yellow. In the centre a shield. A ship at sea, backed by a rock and stars above. The shield surrounded by palm branches and surmounted by a blue motto band with the motto: *In deo conservatio mea.*

11B. Medium yellow. As described by Wissel, but below the arm four hillocks and a tree growing on the one on the extreme right; within a white motto band with the motto: *Idem pacis mediumque velli.*

13A. Blue. White shield surrounded by laurel and palm branches and surmounted by a crown. A red motto band twists across the white ground in the form of a triangle charged with green wreaths and a crown; the motto appears on the base of the triangle, placed uppermost, and is *Præmia sperata.*

13B. An arch with two laurel wreaths; in the opening an altar of variegated marble on which lies a drawn sword. Motto below: *Patriæ vindex et ultor.*

The corners of the flags bore the Royal cypher.

Each squadron of the horsed Regiments had a standard. According to v. Wissel these bore the following emblems:

HANOVERIAN REGIMENTS

LIFE-GUARD REGIMENT

King's Colour—white with gold braid.

a) The arms of England with garter and motto.
Motto below: *Dieu et mon droit.*

b) Entwined G.R.

2. Standard:—white with gold braid.

a) Round orange coloured shield within a garter with the white rose in the centre. Above the shield a helmet with ermine mantling. Motto below: *Nec aspera terrent.*

b) A silver column with fluting of gold; at its base the initials GR; beneath the capital a drawn sword; against the column golden scales. Motto below: *Pro lege et grege.*

Trumpet banner:—white with gold fringes and bearing the Arms of England; trophies of arms in the lower corners, in the upper ones shields with GR surrounded by garters and motto.

VON SCKÖLLN REGIMENT 1B

King's Colour:—white with silver embroidery; on both sides the Arms of England and garter. Motto below: *Dieu et mon droit.*

2. Standard:—yellow richly embroidered with gold.

a) The Arms of England.

b) A silver column, the shaft entwined with laurel wreath; upon the column a gold crown. Motto above: *Mereamur.* On both sides of the column trophies of ancient and contemporary warfare.

This standard bore a silver plate on the staff in memory of 30.8.62 when Trooper Heinrich Grass retrieved the standard after it had fallen into the hands of the French at Johannisberge.

Trumpet banners:—yellow embroidered in gold; in the centre the Arms of England resting on a trophy of arms; on both sides trophies beneath which are slaves in golden chains.

VON DACHENHAUSEN REGIMENT 2A

King's Colour:—white.

a) GR within a garter; motto below: *Dieu et mon droit.*

b) St. George and the dragon; motto above: *Virtus animi superat omnia* (Fig. 220).

2. Standard:—green.
   a) White horse within a garter. Motto below: *Nec aspera terrent.*
   b) Figure of Fame embroidered in gold. Motto above: *Arma virosque cano.*

   Trumpet banners:—green, with horse rampant, surmounted by Elector's crown. Motto below: *Nec aspera terrent.*

FIG. 220. Standard, 2A Regiment of Cavalry.

VON HAMMERSTEIN REGIMENT 2B

King's colour:—white.
   a) A horse on a red ground within a garter. Motto below: *Dieu et mon droit.*
   b) Column with golden scales suspended; on the scales a drawn sword; below the column a trophy of arms. Motto below: *Pro lege et grege.*

2. Standard:—white, like the first standard.

   Trumpet banners:—horse on red ground within a garter. Motto below: *Nec aspera terrent.*

VON GROTTHAUS REGIMENT 3A

King's Colour:—white.
   a) The arms of England within a garter with motto: Motto below: *Dieu et mon droit.*
   b) GR.

2. Standard:—red.
   a) A white horse.
   b) A column with scales hanging unbalanced; upon the column a drawn sword; below it a trophy of arms. Motto above: *Pro lege et grege.*

Trumpet banners:—red with Arms of England and motto below: *Dieu et mon droit.*

VON HODENBERG REGIMENT 3B

King's Colour:—white.

a) The Arms of England with garter and motto *Hony soit qui mal y pense.*

b) GR.

2. Standard:—red.

a) White horse on a red ground within a garter.

b) A column resting on trophy of arms, with scales in balanced position; upon the column a drawn sword. Motto above: *Pro lege et grege.*

Trumpet banners:—red, with the Arms of England and motto.

VON REDEN REGIMENT 4A

King's Colour:—white.

a) Rose on a red ground, within a garter.

b) Column surmounted by a crown. Motto above: *Muoro dove m'attacco.*

2. Standard:—blue.

a) GR embroidered in gold within a garter.

b) Column surmounted by crown. Motto above: *Muoro dove m'attacco.*

Trumpet banner:—white; rose on a red ground within a garter. Motto below: *Nec aspera terrent.*

VON GILTEN REGIMENT 4B

King's Colour:—white.

a) White rose on a red ground within a garter. Motto below: *Nec aspera terrent.*

2. Standard:—colour unknown, probably blue.

a) Similar to the corresponding side of the King's Colour.

b) A lion rampant holding a sword. Motto below: *In pace ad bellum paratus.*

Trumpet banner:—white, white rose on red ground within a garter. Motto: *Nec aspera terrent.*

REGIMENTS OF DRAGOONS

5TH DRAGOONS VON DACHENHAUSEN REGIMENT (IN 1761 VON BELTHEIM)

In the case of Dragoon regiments each of the four squadrons had a standard. The King's colour of this Regiment was white, and had the Arms of England on both sides, borne by the Lion and Unicorn, with garter and motto. The second standard was crimson; one side bore a white horse rampant on red ground within a garter, supported by Lion and Unicorn and below the motto *Dieu et mon droit;* the other side an upright tilting lance with silver pennant on which was represented a similar lance; across the middle of the lance a sheaf of arrows, and above the motto *Virtus unione invicta.*

The third standard was also crimson; one side bore the similar device of the rose, the other a lion couchant, his forepaws resting on a trophy of arms and two trumpets. Motto above: *Parcere subjectis.*

The fourth standard, also crimson, bore on one side the same device as the second and third, and on the other side St. George piercing the dragon with his lance. Motto above: *Virtus animi superat omnia.*

The trumpet banners were crimson with the white rose emblem on red ground within a garter. Motto below: *Nec aspera terrent.*

6TH DRAGOONS, VON BREIDENBACH REGIMENT (IN 1761 G. VON WALTHAUSEN REGIMENT)

The King's colour of this Regiment was white, embroidered in gold. On one side it bore the Arms of England with garter and motto, and on the other a lion couchant on an emblem of victory, with a sword entwined with laurel. Motto above: *Parta tueri.*

The second Standard was blue.

a) GR with crown, within a garter. Motto below: *Dieu et mon droit.*
b) Arm with drawn sword issuing from clouds, the sword entwined with laurel. Motto below: *Cum Deo et gladio.*

The third Standard was also blue.

a) The Arms of England and Lüneburg joined by two intertwining palm trees and a chain. Motto below: *Et stirps et amor jungebant.*

b) A column resting on trophy of arms; at the base of the column GR, resting on it a sword drawn with scales. Motto above: *Pro lege et grege.*

Fourth Standard blue.

a) The Arms of England with motto.

b) St. George on a white horse piercing the dragon lying at the horses feet. Motto above: *Exempla majorum.*

The design of the trumpet banners is not known.

7TH DRAGOONS, VON DER BUSSCHE REGIMENT (1761 VON MULLER REGIMENT

The King's Colour was white, the other three blue.

King's Colour:—

a) White horse on a red ground within a garter; a crown above. Motto below: *Nec aspera terrent.* In the corners entwined GR.

b) Representation of double-headed Janus; on each side trophy of arms. Motto above: *Decet et ornat.*

2. Standard:—

a) Mars above trophy of arms; in one hand a sword in the other a laurel branch. Motto below: *Post bellum nomen.*

b) St. George slaying the dragon. Motto above: *Virtus animi superat omnia.*

3. Standard:—

a) GR beneath a crown; lion and unicorn on either side. Motto below: *Dieu et mon droit.*

b) A golden oriole holding a ring in its beak. Motto below: *Felici ausbicio.*

4. Standard:—

a) Youth, and eagle flying towards the sun. Motto below: *Prosperos favet.*

b) Shield with four crowns; a lion standing over it with a sword. In the corners GR.

Trumpet banners:—blue; horse rampant surmounted by Elector's crown. Motto below: *Nec aspera terrent.*

8TH DRAGOONS, BOCK VON WULFINGEN REGIMENT

The King's Colour of the regiment was white, the others crimson, King's Colour:—

a) The Arms of England with garter and motto.

b) White rose. Motto above: *Quo fas et gloria ducunt.*

2. Standard:—

a) Bare arm with drawn sword issuing from a cloud. Motto above: *Dextra Dei hostes opprimit.*

b) Rose on a red ground within a garter, supported by lion and unicorn. Motto below: *Nec aspera terrent.*

3. Standard:—

a) Large representation of sun with rays, towards which an eagle is flying. Motto above: *Exemplo majorum.*

b) Similar to second standard.

4. Standard:—

a) A raging lion in gold. Motto above: *Pericula non metuit.*

b) Similar to second standard.

Trumpet banners:—red, Arms of England supported, garter and motto below: *Dieu et mon droit.*

BIBLIOGRAPHY

History of the Royal Regiment of Artillery. COLONEL FRANCIS DUNCAN.
Notes on the Early History of the Royal Regiment of Artillery. COLONEL G. CLEAVELAND.
History of the Dress of the Royal Regiment of Artillery. CAPTAIN MACDONALD.
A Short History of the Royal Regiment of Artillery. J. LOUGHLIN.
History of the Corps of Royal Engineers. MAJOR-GENERAL WHITWORTH-PORTER.
History of the Honourable Artillery Company. LT.-COLONEL G. A. RAIKES.
History of the Honourable Artillery Company. MAJOR G. GOULD WALKER, D.S.O., M.C.
Series of Articles on the London Train Bands. MAJOR G. GOULD WALKER, D.S.O., M.C.
Records of the Militia of the County of Southampton. COLONEL LLOYD-VERNEY and LT.-COLONEL J. MOUAT F. HUAT.
A Plan of Discipline for the Use of the Militia of Norfolk. WILLIAM WINDHAM and VISCOUNT GEORGE TOWNSHEND (1759).
The Constitutional Force. COLONEL G. JACKSON-HAY, C.B., C.M.G.
The Journal of the Army Historical Research Society.
Colours and Standards. S. M. MILNE.
A History of the Volunteer Forces. CECIL SEBAG-MONTEFIORE.
History of the British Army. SIR JOHN FORTESCUE.
History of the British Army. SIR SIBBALD SCOTT.
Military Antiquities. GROSE.
Marlborough and the Rise of the British Army. C. T. ATKINSON.
History of the Coldstream Guards. COLONEL MACKINNON.
History of the Scots Guards. MAJOR-GENERAL SIR FREDERICK MAURICE, K.C.M.G., C.B.
History of the Ordnance Services. GENERAL FORBES.
Britain's Sea Soldiers. COLONEL C. FIELD.
John Manners, Marquess of Granby. WALTER EVELYN MANNERS.
Historical Journal of the Campaigns 1757–58–59–60. CAPTAIN JOHN KNOX.
The Great Clan Tartan Myth. LT.-COLONEL M. M. HALDANE. (Published in *The Scots Magazine.*)
The Military History of Perthshire. THE MARCHIONESS OF TULLIBARDINE (now THE DUCHESS OF ATHOLL, D.B.E.).
The Rise of Military Music. H. G. FARMER.
The Life of a Regiment. LT.-COLONEL C. G. GARDYNE.
Ancient Scottish Weapons. JAMES DRUMMOND, R.S.A.
Report of the Sackville Court Martial—
Cavalry. LT.-COLONEL F. N. MAUDE.
Discipline of Light Horse. HINDE.
Institutions Militaires. DE LA PORTERIE.
L'Art Militaire Francoise (1697). P. GIFFART.
THE HANOVERIAN ARMY:
Nec Aspera Terrent. FREDERICK SCHIRMER.
Das Celler Soldaten Buch. FREDERICK SCHIRMER.
Uniformenkunde. R. KNÖTEL.
Material and Notes on Irish Corps. M. S. D. WESTROPP, M.R.I.A.
Material and Notes on Jersey Militia. N. V. L. RYBOT.
Various Histories of Militia Regiments.

INDEX

INDEX

REGIMENTS

INDEX